NEW ERA ENGLISH FOR CAREERS
新时代职业英语专业篇

水利大类

水利英语
词汇手册

总 主 编： 鲁 昕
副总主编： 闫国华
主　 编： 徐秀玲　许家金
编　 者： 王 伶　徐李荣　许晨琛
　　　　　刘朝霞　邹文鑫

外语教学与研究出版社
FOREIGN LANGUAGE TEACHING AND RESEARCH PRESS
北京 BEIJING

图书在版编目（CIP）数据

水利英语词汇手册 / 徐秀玲，许家金主编；王伶等编. -- 北京：外语教学与研究出版社，2022.7
（新时代职业英语 / 鲁昕总主编. 专业篇. 水利大类）
ISBN 978-7-5213-3532-3

Ⅰ. ①水… Ⅱ. ①徐… ②许… ③王… Ⅲ. ①水利工程-英语-词汇-高等职业教育-教学参考资料 Ⅳ. ①TV

中国版本图书馆CIP数据核字(2022)第059485号

出 版 人	王 芳
项目负责	李淑静　史丽娜
责任编辑	卫 昱
责任校对	王霄羽
助理编辑	来菁菁
封面设计	孙莉明　锋尚设计
版式设计	孙莉明
出版发行	外语教学与研究出版社
社　　址	北京市西三环北路19号（100089）
网　　址	http://www.fltrp.com
印　　刷	北京九州迅驰传媒文化有限公司
开　　本	889×1194　1/16
印　　张	7.5
版　　次	2022年9月第1版　2022年9月第1次印刷
书　　号	ISBN 978-7-5213-3532-3
定　　价	36.80元

职业教育出版分社：
地　　址：北京市西三环北路19号 外研社大厦 职业教育出版分社（100089）
咨询电话：010-88819475
传　　真：010-88819475
网　　址：http://vep.fltrp.com
电子信箱：vep@fltrp.com
购书电话：010-88819928/9929/9930（邮购部）
购书传真：010-88819428（邮购部）

购书咨询：(010) 88819926　电子邮箱：club@fltrp.com
外研书店：https://waiyants.tmall.com
凡印刷、装订质量问题，请联系我社印制部
联系电话：(010) 61207896　电子邮箱：zhijian@fltrp.com
凡侵权、盗版书籍线索，请联系我社法律事务部
举报电话：(010) 88817519　电子邮箱：banquan@fltrp.com
物料号：335320001

"新时代职业英语"系列教材编写委员会

总主编

鲁 昕 教育部原副部长

副总主编

闫国华 北京外国语大学

成员（按姓氏拼音排序）

陈晓倩	山东服装职业学院	史少杰	教育部学校规划建设发展中心
邓 飞	华南农业大学		
冯 欣	福建医科大学	宋仁福	无锡科技职业学院
冯正斌	长安大学	孙志农	安徽农业大学
高 霄	华北电力大学	唐克胜	深圳职业技术学院
高小姣	河北水利电力学院	王 珍	青岛农业大学
韩喜春	西安开放大学	夏登山	北京外国语大学
柯爱茹	黎明职业大学	欣 羚	西南交通大学
雷振龙	浙江安防职业技术学院	徐秀玲	北京外国语大学
李淑静	外语教学与研究出版社	许家金	北京外国语大学
李 芝	北京林业大学	许 琳	大连海事大学
林 瑛	广西建设职业技术学院	续 芳	天津体育职业学院
刘建珠	深圳职业技术学院	杨登新	山东水利职业学院
刘娟音	河北工业职业技术学院	杨静怡	中国劳动关系学院
刘 军	上海出版印刷高等专科学校	曾范敬	中国人民公安大学
刘祥荣	日照职业技术学院	张彩华	中国农业大学
刘运锋	安徽工程大学	钟 蔚	武汉纺织大学

前 言

开发背景

  党的十九大以来,党中央、国务院推出了一系列职业教育改革发展的重大举措,《国家职业教育改革实施方案》《职业教育提质培优行动计划(2020—2023年)》《关于推动现代职业教育高质量发展的意见》等政策文件陆续出台,从深化改革到提质培优,再到高质量发展,明确了"十四五"期间职业教育改革发展政策框架。在前途广阔、大有可为的职业教育的宽广大路上,职业外语教育同样承载着践行"提质培优行动计划"、推动现代职业教育高质量发展的艰巨任务和光荣使命。

  为满足数字经济时代高等职业教育对英语的新需求,中国职业技术教育学会会长、教育部原副部长鲁昕教授提议,外研社发挥优势,根据高等职业教育的人才培养目标,依据高职各个专业的人才培养方案,对标教育部最新颁布的《高等职业教育专科英语课程标准(2021年版)》(以下简称"新课标")和《职业教育专业目录(2021年)》(以下简称"新目录"),组织开发"新时代职业英语"系列教材,以满足新时期对高职英语教学的新需求。

  "新时代职业英语"是国内首套"1+19"体系、面向高职专科和职业本科的职业英语教材,由鲁昕教授总体策划。鲁昕教授指出,高等职业英语教材要应国家之需,担发展之责。在这一原则的指导下,本系列教材大胆创新,突破单一的英语语言培养目标的条框限制;锐意探索,巧妙设计"语言、理实和育人"三个维度的目标;精益求精,选取语言素材精、信息密度大、主题立意高的教学内容,构建职业英语教材的大格局。

编写依据

坚持育人为本,落实立德树人根本任务

  充分发挥英语课程育人的功能,在话题、选材、语言等各个层次上严格把关,将专业精神、职业精神和工匠精神有机融入教材,实现立德树人教育细致化、过程化,让学生在发展英语语言能力的过程中,能够更好地培育和践行社会主义核心价值观。

突出职教特色,注重质量为先

  遵循技术技能人才成长规律,按照各个专业大类对应的职业岗位对英语的实际需求进行编写,体现英语学科的特点,突出职业教育特色,实现教材的职业性、实用性、适用性。

坚持产教融合,校企"双元"开发

  教材编写团队由英语和相关专业的专家、教科研人员、一线教师、行业企业技术人员和能工巧匠等构成,教材语言准确,内容紧跟产业发展趋势和人才培养需求,强化学生实习实训。

充分利用现代信息技术,创新教材形式

  "新时代职业英语"系列教材充分利用各种信息技术手段提升学生的学习效率和教师的教学效果。系列教材配有试题库、数字课程、助教课件等,将信息技术和英语教学充分融合。

教材体系

  "新时代职业英语"系列教材包括通用篇和专业篇两部分。

通用篇

  通用篇覆盖高职专科和职业本科两个学段。其中,《通用英语》

（共 2 册）可作为高职公共英语基础课教材使用，《通用英语（本科版）》（共 2 册）可作为职业本科或应用型本科的大学英语基础课教材使用。通用篇由学生用书、评估手册和教师用书组成，并配有 iTest 试题库、U 校园数字课程、随身学数字课程、电子教案、PPT 课件、TOP 课件、单元测试卷、期中和期末试卷等数字资源。

专业篇

专业篇分为"1+19"大类。其中，"1"是指作为通识课教材的《人工智能英语》，"19"是指高职 19 个专业大类的专业英语教材。"1+19"的每个类别都包括学习手册、词汇手册和使用手册。专业篇可作为高职专科、职业本科以及应用型本科的专业英语课教材使用。

1. 学习手册

由于各个专业的人才培养方案以及对应的典型工作岗位对英语的实际需求不尽相同，每个专业大类的专业英语教材的编写模式各有特点。现代服务业，如酒店、旅游、物流等行业的专业英语教材，以该行业的工作过程、典型工作环节和场景为参照来组织内容，根据主要工作任务所需的英语知识和技能设计英语学习任务，培养学生用英语处理相关业务的能力。先进制造业，如电子信息、交通运输等行业的专业英语教材，主要以专业文本、产品说明书的阅读为主，侧重培养学生的专业英语阅读能力。

2. 词汇手册

专业词汇是行业英语教学的难点。词汇手册基于语料库编写，采用每日一词的形式筛选出该行业最常用的 365 个词，并提供该词最常见的搭配和使用场景。词汇手册配有词汇学习"U 词"App，具有词汇学习、复习以及中英文词典查询的功能，将信息技术应用和英语教学相结合，便于学生随时学习。

3. 使用手册

电子版使用手册为教师提供丰富的教学参考，包括参考答案、课文译文、音频脚本、补充资源、教学建议、教学步骤、拓展活动等，切实帮助教师提高教学的效率和效果。

教材特色

新理念：语言、理实、育人有机融合

"新时代职业英语"秉承"以语言为基础，以理实为路径，以育人为目标"的理念，将语言知识和语言技能有机融合，构建教材之基；将专业理论与使用说明对比呈现，构架教学之径；将文化自信与工匠精神继承发展，构筑教育之魂。

新标准：紧扣"新课标"

"新时代职业英语"包括通用篇和专业篇，与"新课标"课程结构要求的基础模块和拓展模块高度匹配。通用篇对应基础模块的职场通用英语课程，严格按照"新课标"课程内容中对主题类别、语篇类型、语言知识、文化知识、职业英语技能、语言学习策略的具体要求编写，旨在培养学生的英语应用能力，引领学生在四项英语学科核心素养方面融合发展。专业篇"1+19"大类对应拓展模块的课程要求，其中"1"《人工智能英语》是专业通识课教材，对应素养提升英语课程；19个专业大类英语教材对应职业提升英语课程，旨在提升学生的职业素养，为其进入职场作好准备。

新体系：贴合"新目录"

"新时代职业英语"是国内首套"1+19"体系的职业英语教材。其中，"19"对应"新目录"中的19个高等职业教育专科专业，旨在帮助进入不同工作岗位的各专业学生积累专业英语词汇，满足其在职场中的涉外沟通需要，全方位服务国家战略和经济社会发展需求。

新逻辑：思政融入、理实一体

通用篇特设中华文化板块，呈现优秀文化，弘扬时代风尚，以突出文化自信为宗旨，将思政教学真正融入到英语语言学习中，体现中华优秀文化，全面落实课程思政要求。专业篇学习手册创新性引入"说明书"文体，创建"理实一体"的特色课堂；词汇手册采用语料库技术遴选高频词，配套的"U词"App利用记忆规律和大数据分析辅助学习，让学习进入智能时代。

新载体：可听、可视、可练、可互动

提供配套的音频、视频、试题库、数字课程等全方位立体化数字资源，旨在打造一套可听、可视、可练、可互动的融媒体教材，助力教师实现线上线下混合模式教学。

编写团队

中国职业技术教育学会会长、教育部原副部长鲁昕教授担任"新时代职业英语"系列教材的总主编和总策划，北京外国语大学原副校长闫国华教授担任副总主编。本系列教材由北京外国语大学牵头，以中国职业外语教育发展研究中心成员为主要研发力量，同时汇聚了全国各地本科和高职院校的专家和教师，各行业企业也广泛参与，出谋划策。我们相信，凝聚多方经验与智慧的"新时代职业英语"系列教材必将为我国职业教育外语教学改革助力，为我国职业教育发展赋能。

编写说明

《水利英语词汇手册》是"新时代职业英语专业篇"的一个分册,可供高职专科和职业本科院校"水利大类"相关专业学生使用,旨在帮助学生积累水利专业英语词汇,培养学生用英语处理水利行业相关业务的能力。本书收录的词汇涉及水利领域多种职业场景,可满足水利行业从业人员工作过程中的基本英语词汇量需要,并为学生因职业发展而进一步学习专业英语打好基础。

一、编写理念

学以致用是职业院校学生学习专业英语的第一要义。然而,要让静态词语变为活用语汇,绝非易事。《水利英语词汇手册》采用"四用原则"编写,即"真材实用,优选常用,单词连用,情境活用",希望为学生提供选材地道、语境典型的优质语言材料。

二、内容构成

本书收录核心词365个,旨在帮助学生利用碎片化的时间每天学习一个词条。每个词条呈现的内容主要包括音标、词性、词义、常见短语和释义、英文例句和译文,部分词条还提供扩展知识、词语辨析、相关概念等注释。除此之外,本书还有音标表和总词表两个附录。

· 词条遴选

本书中提供的词条、短语和例句均甄选自水利行业真实场景的英语语料库。所有词条的选定采用统计方法,按常用程度高低排序,并以星号(★)表示频度:五星词(★★★★★)为极高频词,四星词(★★★★)为高频词,三星词(★★★)为普通高频词。

全书共收录核心词 365 个，其中，五星词 100 个，四星词 100 个，三星词 165 个。

- **示例呈现**

　　词语用法示例采用两种形式：短语和句子。单词词义往往难以独立确定，将其放在短语、句子等真实语境中，意义可以更加完整和明晰地呈现。

- **趣味阅读**

　　为了更直观生动地呈现词条释义，让词汇学习不再是枯燥的死记硬背，本书还配有图片和注释等丰富的内容，帮助学生快乐阅读，轻松记忆。

三、编写团队

　　"新时代职业英语"系列教材专业篇中词汇手册的总策划为北京外国语大学中国外语与教育研究中心、人工智能与人类语言重点实验室的许家金教授。

　　《水利英语词汇手册》的主编为徐秀玲和许家金，编者为王伶、徐李荣、许晨琛、刘朝霞和邹文鑫。本书是国家社科基金项目"概率语境共选视角下的多语外汉词典数据库建设与研究"（21BYY021）的阶段性成果。

　　本书凝聚了众人的智慧和心力，但难免存在错误和不足，请广大师生批评指正，以便我们不断改进。

范例说明

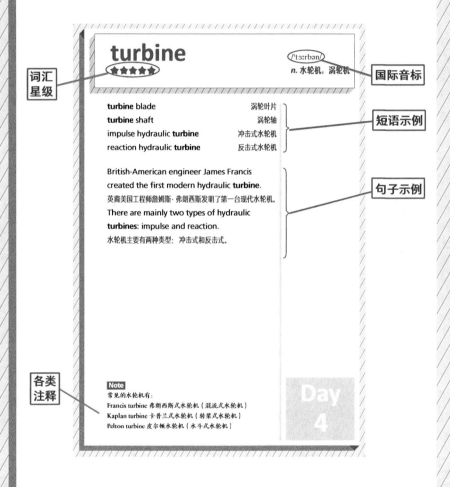

目 录

第一部分　五星词汇……………………………………………………1

第二部分　四星词汇……………………………………………………103

第三部分　三星词汇……………………………………………………159

附录一　音标表…………………………………………………………215

附录二　总词表…………………………………………………………216

第一部分

五星词汇

water
★★★★★

/ˈwɔːtər/

n. 水，水域

water pump	水泵
water quality	水质
water resources	水资源
water right	水权
water supply	给水，供水

This district faces a serious **water** quality challenge.
这个地区面临着严重的水质挑战。

Management of **water** resources is a complex decision-making process.
水资源管理是一个复杂的决策过程。

Clean **water** is critical to plants and animals that live in it.
洁净的水对于水生动植物至关重要。

Day 1

hydraulic
★★★★★

/haɪˈdrɒːlɪk/

adj. 水力学的，液压的，水下硬化的

hydraulic cement	水硬水泥
hydraulic engineering	水利工程
hydraulic jump	水跃
hydraulic structure	水工建筑物
hydraulic system	液压系统

Hydraulic engineering often plays a key role in designing and building tunnels, bridges and highways.
设计和建造隧道、桥梁以及高速公路时，水利工程往往起着关键作用。

Hydraulic systems are used in all kinds of large and small industrial settings.
液压系统在各类大大小小的工业场景中都有应用。

Day 2

soil
★★★★★

/sɔɪl/

n. 土壤

soil and water conservation　　　水土保持
soil erosion　　　土壤侵蚀
soil moisture　　　土壤湿度
soil surface　　　土表

Soil is an important type of natural resources, just as the air and water that surround us are.
土壤是一种重要的自然资源，正如我们周围的空气和水一样。
A system of **soil** sensors allows farmers to monitor **soil** conditions on a daily basis.
土壤传感器系统让农民可以每天监测土壤状况。

Day 3

flow

★★★★★

/floʊ/

n. 水流，流量

base **flow**	基流
groundwater **flow**	地下水流
peak **flow**	洪峰流量
river **flow**	河川径流
water **flow**	水流

In most streams, base **flow** comes largely from groundwater.
大多数河流的基流主要来源于地下水。

The water **flow** is blocked by a dam, allowing an artificial lake to be formed.
水流被大坝阻断，形成一个人工湖。

Typical **flow** rate units are liters per second or cubic meters per second.
流量的单位通常是升每秒或立方米每秒。

Day 4

turbine
★★★★★

/ˈtɜːrbən/
n. 水轮机，涡轮机

turbine blade　　　　　　　　涡轮叶片
turbine shaft　　　　　　　　涡轮轴
impulse hydraulic **turbine**　　冲击式水轮机
reaction hydraulic **turbine**　　反击式水轮机

British-American engineer James Francis created the first modern hydraulic **turbine**.
英裔美国工程师詹姆斯·弗朗西斯发明了第一台现代水轮机。
There are mainly two types of hydraulic **turbines**: impulse and reaction.
水轮机主要有两种类型：冲击式和反击式。

Note
常见的水轮机有：
Francis turbine 弗朗西斯式水轮机（混流式水轮机）
Kaplan turbine 卡普兰式水轮机（转桨式水轮机）
Pelton turbine 皮尔顿水轮机（水斗式水轮机）

Day 5

structure

★★★★★

/ˈstrʌktʃər/

n. 结构，建筑物

structure drawing	结构图，构造图
concrete **structure**	混凝土结构
drop **structure**	落差建筑物
intake **structure**	进水建筑物
steel **structure**	钢结构

A hydraulic **structure** can be built in a river, a sea, or any body of water.
水工建筑物可建于河流、海洋或其他任何水体中。

The water control **structures** should be built before letting any water flow through the canal.
在让水流入运河之前，应该建造控水结构。

A dam is a large, man-made **structure** built to contain some water bodies.
大坝是大型的人造建筑物，用来拦截某些水体。

dam
★★★★★

/dæm/

n. 坝，水坝

dam site	坝址
arch **dam**	拱坝
concrete **dam**	混凝土坝
embankment **dam**	土石坝，填筑坝

A **dam** is built on a soil or rock foundation.
大坝修建在土地基或岩石地基上。

A **dam** with a large drop is constructed to raise the potential energy of water.
修建落差较大的大坝是为了增加水的势能。

Dams are generally constructed to serve purposes such as the generation of hydroelectric power, to control flooding, to reduce the imminence of drought and to develop tourism.
建水坝通常是为了水力发电、控制洪水、缓解干旱和发展旅游业等。

Day 7

equipment

★★★★★

/ɪˈkwɪpmənt/

n. 设备

equipment selection	设备选型
drip irrigation **equipment**	滴灌设备
electromechanical **equipment**	机电设备
heavy **equipment**	重型设备
mechanical **equipment**	机械设备

This type of heavy **equipment** is used for underwater excavations.
这类重型设备用于水下挖掘。

Abnormal conditions of this **equipment** need to be alarmed.
此设备如出现异常，需有报警提示。

The **equipment** is portable, simple to operate, and includes rechargeable battery.
这个设备便于携带，操作简单，还配有充电电池。

river

/ˈrɪvər/

n. 河流

★★★★★

river bed	河床
river flow	河流径流
river regulation	河道整治
river valley	河谷
canalized **river**	渠化河道
natural **river**	天然河流

The impact that power plants and reservoirs have on **river** basins is undeniable.
发电厂和水库对河流流域的影响是毋庸置疑的。

Rivers are a natural waterway which can be used as a means of transport.
河流是一种天然航道,可作为一种交通方式。

Day 9

design
★★★★★

/dɪˈzaɪn/

n. 设计

design flood	设计洪水
design phase	设计阶段
computer aided **design**	计算机辅助设计
engineering **design**	工程设计
system **design**	系统设计

The source of water has a major effect on water distribution system **design**.
水源对配水系统的设计有着重要影响。

The initial phases of surveys and **design** may extend over months.
调查和设计的初始阶段可能会持续几个月。

drainage
★★★★★

/ˈdreɪnɪdʒ/
n. 排水

drainage area   排水区域
drainage ditch   排水沟
drainage system   排水系统
subsurface **drainage**   地下排水
surface **drainage**   地面排水

Such **drainage** systems not only help in preventing floods, but also improve water quality.
这种排水系统不仅有助于防洪，还能改善水质。
The removal of free water by **drainage** allows soil to warm up quickly.
通过排水去除自由水可以使土壤迅速升温。

Day 11

plant
★★★★★

/plænt/

n. 工厂

hydropower **plant**	水电厂，水电站
pumping **plant**	泵站
water treatment **plant**	净水厂，给水处理厂

At hydropower **plants**, water flows through a pipe or penstock, then pushes against and turns blades in a turbine to spin a generator to produce electricity.
在水电厂，水流经管道或压力管道，推动涡轮机叶片旋转，从而使发电机发电。

A coal-fired power **plant** uses steam to turn the turbine blades; whereas a hydropower **plant** uses falling water to turn the turbine.
燃煤发电厂使用蒸汽来推动涡轮叶片；而水电厂则利用落下的水流来转动涡轮机。

Day 12

run-off
★★★★★

/ˈrʌn ɔːf/
n. 径流

flood **run-off**	洪水径流
spring **run-off**	春季径流
storm **run-off**	暴雨径流
surface **run-off**	地表径流

It is difficult to predict exactly when the spring **run-off** starts and finishes.
很难准确预测春季径流何时开始、何时结束。

Surface **run-off** is the primary cause of soil erosion.
地表径流是土壤侵蚀的主要原因。

Run-off can come from both natural processes and human activity; the most familiar type of natural **run-off** is snowmelt.
径流可以来源于自然过程，也可以由人类活动产生。最常见的自然径流是雪融水。

Day 13

construction

★★★★★

/kənˈstrʌkʃən/

n. 建造，建筑物

construction cost	工程造价
construction diversion	施工导流
construction industry	建筑行业
construction material	建筑材料
construction risk	施工风险
construction schedule	施工进度计划
construction specification	施工规范

Permits are only authorized if protection measures are involved in the **construction** works.
建筑工程必须包含安全防护措施，才会获发许可证。
The **construction** management plan should include relevant policies that guide the work.
施工管理方案应包含指导施工的相关政策。

Day 14

irrigation
★★★★★

/ˌɪrəˈɡeɪʃən/
n. 灌溉

irrigation canal 灌溉渠道
irrigation system 灌溉系统
furrow **irrigation** 沟灌
sprinkler **irrigation** 喷灌
surface **irrigation** 地面灌溉

In certain cases, it is allowable for rainwater to flow into an **irrigation** canal.
在某些情况下，可以允许雨水进入灌溉渠道。
It's generally believed that the biggest advantage of drip **irrigation** technology is water saving.
大家普遍认为，滴灌技术最大的优点是节水。

Day 15

pump
★★★★★

/pʌmp/

n. 泵，抽水机
v. 从地下抽出（水、油）等

pump station	泵站
pumped storage	抽水蓄能
centrifugal **pump**	离心泵
hand **pump**	手摇泵
hydraulic **pump**	液压泵，水力泵
well **pump**	井泵

An old-fashioned hand **pump** requires no electricity.
老式的手摇泵不需要用电。

There are two common types of water **pumps**: vane and positive displacement.
水泵有两种常见类型：叶片式和容积式。

The centrifugal **pump** can be defined as a hydraulic machine which converts mechanical energy into hydraulic energy by means of centrifugal force acting on the fluid.
离心泵是一种液压机，通过作用在流体上的离心力将机械能转化为水能。

Day 16

supply
★★★★★

/sə'plaɪ/
n. 供给，补给

municipal water **supply**	城市供水
potable water **supply**	饮水供给
power **supply**	供电
public water **supply**	公共供水
water **supply** system	给水系统，供水系统

The evolution of public water **supply** systems is tied directly to the growth of cities.
公共供水系统的发展与城市的发展直接相关。

Water **supply** to rural areas can be sourced from rainwater, groundwater and surface water.
农村地区的水供给可以来自雨水、地下水和地表水。

Among various sources of fresh water **supply**, groundwater is by far the most practicable choice.
各种淡水供给源中，地下水是迄今为止最可行的选择。

Day 17

hydropower
★★★★★

/ˈhaɪdroʊˌpaʊr/
n. 水力发电，水电

hydropower facilities　　　　　水电设施
hydropower project　　　　　　水电项目
hydropower station　　　　　　水电站
small **hydropower**　　　　　　　小水电

Small and micro **hydropower** stations utilize water from rivers and avoid big environmental impacts.
小型和微型水电站利用河流水，避免对环境产生巨大影响。
Small **hydropower** potential is given in hilly or mountainous regions.
山地丘陵地区的水电蕴藏量较小。

Day 18

flood
★★★★★

/flʌd/

n. 洪水

flood bypass　　　　　　　　分洪道
flood control　　　　　　　　防洪
flood plain　　　　　　　洪泛平原，河漫滩
flood routing　　　　　　　　洪水演进
flash **flood**　　　　　　　骤发洪水，暴洪

A dike is a structure of **flood** protection.
堤坝是一种防洪建筑物。
The risk of a major spring **flood** could increase if heavy winter precipitation occurs.
如果冬季有强降水，发生较大春汛的风险会增加。

Day 19

project
★★★★★

/ˈprɑːdʒekt/

n. 工程，项目

project cost	项目成本，工程造价
project development	项目开发
project implementation	项目实施，项目执行
project management	项目管理
construction **project**	建设工程项目

A good detailed **project** report would lead to smooth implementation of a **project**.
一份优秀详细的项目报告会促使项目顺利进行。

The integrated **project** management plan should involve major **project** components.
项目综合管理计划应包含主要的项目组成部分。

Day 20

generator

★★★★★

/ˈdʒenəreɪtər/

n. 发电机

generator room	发电机房
generator shaft	发电机轴
generator stator	发电机定子
induction **generator**	感应发电机
synchronous **generator**	同步发电机

The hydraulic **generator** or generating unit shall be installed according to the manufacturer's instructions.
水轮发电机或发电机组应按照制造商的说明安装。

A **generator** is a piece of equipment that converts mechanical energy into electrical energy.
发电机是把机械能转换成电能的设备。

Day 21

pipe
★★★★★

/paɪp/

n. 管，管道

pipe diameter	管径
pipe network	管网
drip irrigation **pipe**	滴灌管
hydraulic **pipe**	液压管道

The drip irrigation **pipe** is a low-pressure **pipe** system.
滴灌管是一种低压管道系统。

The steel **pipe** is very strong, but has a shorter service life than the concrete pipe.
钢管很结实，但是使用寿命比混凝土管短。

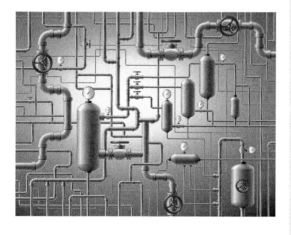

Day 22

control
★★★★★

/kən'troʊl/

n. 控制，管理

control board	控制板
control room	控制室
control system	控制系统
control valve	控制阀
remote **control**	遥控，远程控制
supervisory **control**	监督控制

Control signals are signals given from the **control** board to various pieces of equipment.
控制信号是指从控制板发送到各种设备的信号。

When the **control** gate is fully opened, the maximum amount of water will be released through the penstock.
控制闸门完全打开时，压力管道中会释放出最大水量。

Day 23

engineering
★★★★★

/ˌendʒəˈnɪrɪŋ/
n. 工程（设计）

engineering geology	工程地质学
engineering hydrology	工程水文学
engineering structure	工程结构
port **engineering**	港口工程

Environmental **engineering** focuses on protecting the environment by reducing waste and pollution.
环境工程重点关注如何通过减少浪费和污染来保护环境。

Environmental **engineering** utilizes engineering techniques and principles of related disciplines to improve environmental quality.
环境工程学运用工程技术和有关学科的原理来改善环境质量。

Civil **engineering** concerns the planning, building, and repair of roads, bridges, large buildings, etc.
土木工程关注道路、桥梁、大型建筑等的规划、建筑以及养护。

Day 24

quality
★★★★★

/ˈkwɑːləti/

n. 质量，品质

quality assurance — 质量保证
quality control — 质量控制
air **quality** — 空气质量
environmental **quality** — 环境质量

The water **quality** inspector samples the water distribution system from a variety of locations throughout the city.
水质检测员在全城的不同地点对配水系统进行取样。

In areas with high **quality** water sources, water charges will be much lower.
在拥有优质水源的地区，水费会低很多。

Day 25

reservoir
★★★★★

/ˈrezərvwɑːr/
n. 水库，蓄水池

reservoir regulation	水库调度
reservoir seepage	水库渗漏
lower **reservoir**	下水库
upper **reservoir**	上水库

A **reservoir** is built for water storage.
水库是为蓄水而建的。

Hydrologists use topographic maps and aerial photographs to calculate **reservoir** depths and storage capacity.
水文工作者用地形图和航拍照片来计算水库深度和储水量。

Day 26

surface
★★★★★

/ˈsɜːrfɪs/

n. 表面，表层

surface feature	地形，地貌
surface tension	表面张力
surface water	地表水
ground **surface**	地面，地表

A huge 70% of the Earth's **surface** is covered with water.
地球表面 70% 之多的面积被水覆盖。

The tunnel was some 91 m below the **surface**.
隧道在地表约 91 米下。

The liquid or solid water in the cloud, such as rain, falls on the Earth's **surface**, and this is called precipitation.
云里的液态水或固态水，比如雨水，落在地球表面，这个过程叫作降水。

Day 27

resource

★★★★★

/ˈriːsɔːrs/

n. 资源

resource planning	资源规划
financial **resources**	财政资源
hydropower **resources**	水能资源
scarce **resources**	稀缺资源
water **resources** management	水资源管理

Visual interpretation has served as a major tool for assessing groundwater **resources**.
目视判读是评估地下水资源的主要方法。

Fresh water is a finite and vulnerable **resource**.
淡水是一种有限且脆弱的资源。

Freshwater lakes and shallow groundwater represent an available freshwater **resource** for humans.
淡水湖和浅层地下水是人类可用的淡水资源。

Day 28

canal
★★★★★

/kəˈnæl/

n. 渠道，运河

canal barge 运河驳船
artificial **canal** 人工渠道
inland **canal** 内陆运河

Under gravity irrigation, water is distributed by means of open **canals**.
在自流灌溉下，水通过明渠进行分配。

The Panama **Canal** was originally administered and supervised by the military.
巴拿马运河最初由军队管理监督。

Day 29

discharge
★★★★★

/ˈdɪstʃɑːrdʒ/ *n.* 排放，流量
/dɪsˈtʃɑːrdʒ/ *v.* 排放

discharge capacity　　　　　泄流能力
discharge coefficient　　　　流量系数

The peak **discharge** of a flood is influenced by many factors, including the intensity and duration of storms and snowmelt, the topography and geology of the river basin and vegetation.
洪峰流量受许多因素影响，包括风暴和融雪的强度和持续时间、流域的地形地质以及植被。

Some jurisdictions require that storm water should be processed at certain degree before being **discharged** directly into sewers.
一些辖区规定暴雨积水在直接排入下水道之前要经过一定程度的处理。

Day 30

groundwater
★★★★★

/ˈɡraʊndˌwɔːtər/

n. 地下水

groundwater contamination 地下水污染
groundwater recharge 地下水补给
groundwater resources 地下水资源

Under the pull of gravity, **groundwater** flows slowly and steadily through the aquifer.
在重力作用下，地下水缓慢而稳定地流过含水层。
It is useful to know how existing **groundwater** supply could be affected.
了解现有地下水供给可能受到何种影响是有用的。

Day 31

area
★★★★★

/ˈeriə/

n. 面积，区域

area of dissipation	消融区
affected **area**	受灾地区
downstream **area**	下游地区
hilly **area**	丘陵山区
semi-arid **area**	半干旱地区

According to the Third National Land Resource Survey, China has a total **area** of approximately 1.9 billion *mu* in arable lands.
第三次全国国土调查显示，我国耕地总面积约为 19 亿亩。
The geographical information system (GIS) can be used to give statistics on affected **areas**.
地理信息系统（GIS）可以用来提供受灾地区的统计数据。

Day 32

data

/ˈdeɪtə/

n. 数据，资料

★★★★★

data collection	数据采集
data logger	数据记录器
hydrological **data**	水文资料，水文数据
monitoring **data**	监测数据

The satellite provides wireless **data** access for all base stations that request for geospatial **data**.
卫星向所有请求地球空间数据的基站提供无线数据访问权限。

Remote computer terminals can access the system and obtain **data**.
远程计算机终端可以访问该系统并获取数据。

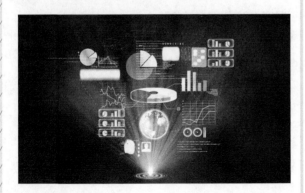

Note
data 的单数形式为 datum，复数形式 data 使用较多。

Day 33

monitoring

/ˈmɑːnɪtərɪŋ/
n. 监测，监控

★★★★★

monitoring program 监控程序
climate **monitoring** 气候监测
remote **monitoring** 远程监控

Water quality **monitoring** reflects the health of surface water bodies as a snapshot in a period (a week, a month or a year).
水质监测以一段时间（一周、一月或一年）内快照的形式反映地表水体的健康状况。

Long-term **monitoring** of water quality is challenging.
水质的长期监测具有挑战性。

Water level **monitoring** plays a critical part in hydrometry.
水位监测在水文测验中起着至关重要的作用。

Day 34

conservation

★★★★★

/ˌkɑːnsərˈveɪʃən/

n. 保护，保持，保存

conservation area	水土保持区
conservation plan	保护计划
conservation tillage	保持耕作

Hydraulic criteria is used to develop **conservation** measures.
根据水力标准来设计保护措施。

These villages are the models in soil and water **conservation** in the region.
这些村庄是该地区水土保持工作的模范。

It is important to encourage environmental **conservation** and awareness.
促进环境保护和提高环境意识是非常重要的。

Day 35

power
★★★★★

/paʊr/

n. 功率，电力，动力

power distribution	配电
power generation	发电
power output	功率输出
power transmission	输电

Hydraulic **power** stations are becoming very popular because the reserves of fuels are depleting day by day.
由于燃料储备日益枯竭，水力发电站变得极受欢迎。

The vehicle had better **power**, better tires, and better brakes.
这种车辆的动力、轮胎和刹车都更好。

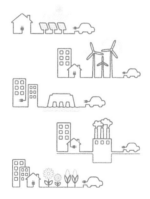

Day 36

watershed
★★★★★

/ˈwɔːtərʃed/
n. 流域，集水区

watershed boundary	流域边界
watershed characteristics	流域特征
watershed management	流域管理
small **watershed**	小流域

Watershed management involves management of the land surface and vegetation so as to conserve and utilize the water that falls on the **watershed**.
流域管理包括地表管理和植被管理，以保存和利用流域内的降水。

The smaller the **watershed** is, the more significant influences individual **watershed** characteristics have on peak discharges.
流域越小，单个流域特征对洪峰流量的影响越显著。

Day 37

concrete
★★★★★

/ˈkɑːŋkriːt/

n. 混凝土

hardened **concrete**   硬化混凝土
precast **concrete**   预制混凝土
reinforced **concrete**   钢筋混凝土

The dam can be made of a number of materials including **concrete**, stone and even earth.
大坝可以用混凝土、石头甚至泥土等多种材料建造。
Hydrophobic cement is used to create **concrete**.
防潮水泥用于制造混凝土。

Day 38

cost
★★★★★

/kɒːst/

n. 成本，费用

cost overrun	成本超支
capital **cost**	资本成本
maintenance **cost**	维护费用，维修费用
operating **cost**	经营成本

The **cost** estimator has determined the basic wages and benefits.
成本预算专员确定了基本工资和福利。

Operation and maintenance **costs** of hydropower account for a small proportion of investment **costs** per year.
水电的运维成本每年占投资成本的比重较小。

Day 39

hydro-

/ˈhaɪdroʊ/

prefix. 水的，使用水的

hydroelectricity　　　　水力发电
hydrometer　　　　　　液体比重计
hydrostatics　　　　　　水静力学
hydrotherapy　　　　　　水疗法

Main inlet valves are a very vital organ of a **hydro**power plant.
主进水阀是水电厂的关键部件。

In a **hydro**static test, the pipes are filled with water and put under pressure to check for leaks.
在静水压实验中，把水装满管道并加压，以检查是否有泄漏。

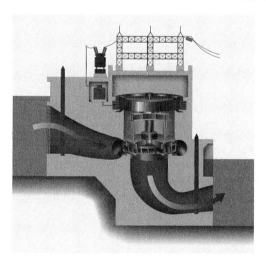

treatment
★★★★★

/ˈtriːtmənt/

n. 处理，治理

biological **treatment**	生物处理
percolation **treatment**	渗漏处理
sewage **treatment** plant	污水处理厂

The sewage **treatment** has three stages, which are primary, secondary, and tertiary **treatment**.
污水处理分三个阶段，即一级、二级和三级处理。

The water **treatment** specialist routinely checks the equipment to make sure it is operating normally.
水处理专家定期检查设备，确保其正常工作。

The government pays close attention to the **treatment** of polluted rivers.
政府密切关注污染河流的治理工作。

Day 41

erosion

★★★★★

/ɪˈroʊʒən/

n. 侵蚀，腐蚀，风化

erosion control	侵蚀防治
coastal **erosion**	海岸线侵蚀
sheet **erosion**	片蚀
wind **erosion**	风蚀

Good crop management reduces soil **erosion** by water and wind.
良好的作物管理可以减少水和风对土壤的侵蚀。

Wind **erosion** is one of the factors leading to the soil degradation and desertification in arid and semi-arid areas of northern China.
风蚀是导致我国北方干旱半干旱地区土壤退化和沙漠化的原因之一。

Day 42

hydrology
★★★★★

/haɪˈdrɑːlədʒi/
n. 水文学

hydrology modernization	水文现代化
groundwater **hydrology**	地下水水文学
surface water **hydrology**	地表水水文学

Basic systems in **hydrology** enable better water management and sustainable development.
水文学的基础系统使更好的用水管理和可持续发展成为可能。

Hydrology modeling cannot be performed effectively without GIS.
没有地理信息系统，就无法有效地进行水文建模。

Hydrology is an extremely important field of study, dealing with one of the most valuable resources on Earth: water.
水文学是一个极其重要的研究领域，研究的是地球上最宝贵的资源之一：水资源。

Day 43

method
★★★★★

/ˈmeθəd/

n. 方法，方式

method of approximation 近似法
disinfection **method** 消毒方法
effective **method** 有效方法
irrigation **method** 灌溉方式

With simple modifications, these traditional **methods** can be made more efficient in operation.
经过简单的改良，这些传统方法在运行中可以更加高效。
Hydropower is one of the cleanest **methods** of power generation.
水力发电是最清洁的发电方式之一。
Some **methods** of flood control have been practiced since ancient times, including planting vegetation, terracing hillsides and constructing floodways.
某些防洪措施从古时候就开始投入实践了，包括植被覆盖、在山腰修梯田和修建泄洪道。

maintenance
★★★★★

/ˈmeɪntənəns/

n. 维修，维护，保养

maintenance engineer	维修工程师
equipment **maintenance**	设备维护
regular **maintenance**	定期保养，定期维修

The **maintenance** crew is responsible for the equipment in the hydropower station.
维修管理员对水电站的设备负责。

This portable weather station requires less **maintenance** than conventional ones.
与传统气象站相比，这种便携式气象站所需的维护更少。

Day 45

unit
★★★★★

/ˈjuːnɪt/

n. 单位，组件，部件

unit price	单价
generating **unit**	发电机组
hydraulic **unit**	液压部件
power **unit**	功率单元

Waste and loss may be included in material **unit** price calculations.
计算材料单价时，可将浪费和损耗包括进来。

The removal rate per **unit** of area varies according to the size of the drainage area.
每单位面积的去除率随流域面积变化而变化。

The installed capacity is measured in **units** of power called "watt".
装机容量是用功率单位"瓦特"来计量的。

Day 46

wastewater

★★★★★

/'weɪstˌwɔːtər/

n. 废水，污水
adj. 废水的，污水的

wastewater discharge	废水排放，污水排放
wastewater treatment	废水处理，污水处理
domestic **wastewater**	生活污水
industrial **wastewater**	工业废水
untreated **wastewater**	未处理的废水

Primary treatment removes most suspended solids from **wastewater**.
一级处理去除废水中大部分悬浮固体。

Wastewater has previously been used and may contain oils, chemicals, food, or sewage.
废水是使用过的水，可能含有油、化学物质、食品或污物。

Day 47

storage

★★★★★

/ˈstɔːrɪdʒ/

n. 贮存，存储

storage capacity	存储容量
storage irrigation	储水灌溉
storage reservoir	蓄水库
pumped **storage** station	抽水蓄能电站
water **storage**	蓄水

The greatest water **storage** lies near the surface.
绝大部分水储存在地表附近。

All data is sent to the cloud for **storage**.
所有数据发送到云端存储。

Day 48

management
★★★★★

/ˈmænɪdʒmənt/
n. 管理

construction **management**	施工管理
risk **management**	风险管理
river **management**	河道管理
solid waste **management**	固体废物管理

The environmental impact assessment (EIA) is a useful component of good environmental **management**.
环境影响评价（EIA）是良好环境管理的有用组成部分。
Water quality monitoring is the foundation of water quality **management**.
水质监测是水质管理的基础。

Day 49

precipitation

★★★★★

/prɪˌsɪpɪˈteɪʃən/
n. 降水，降水量

precipitation record	降水记录
annual **precipitation**	年降水量
normal **precipitation**	正常降水量
spring **precipitation**	春季降水

Precipitation in China generally follows the same pattern as temperatures, decreasing from the southeast to the northwest.
中国的降水分布和气温基本一致，自东南向西北逐渐递减。

On all of our scenarios winter **precipitation** increases substantially.
从各种情况看，冬季的降水量有显著增加。

Day 50

valve
★★★★★

/vælv/

n. 阀

air **valve**	空气阀
butterfly **valve**	蝶形阀
check **valve**	止回阀
hydraulic **valve**	液压阀
inlet **valve**	进口阀
pressure relief **valve**	卸压阀

The **valve** house contains main sluice **valves** and automatic isolating **valves**.
阀室包含主闸阀和自动隔离阀。

Our trained technicians can provide pump and **valve** maintenance.
我方受过培训的技术人员可提供泵和阀门的维护服务。

Day 51

hydroelectric
★★★★★

/ˌhaɪdroʊ-ɪˈlektrɪk/
adj. 水电的，水力发电的

hydroelectric generator	水力发电机
hydroelectric industry	水电行业
hydroelectric plant	水电厂

The presence of **hydroelectric** dams can often change migration patterns and hurt fish populations.
水电站大坝经常会改变洄游模式，使鱼类种群减少。

A **hydroelectric** plant technician may have to work over 40 hours a week.
水电厂的技术人员可能一周会工作 40 个小时以上。

Day 52

material
★★★★★

/məˈtɪriəl/
n. 材料，原料

hazardous **material**   有害材料
organic **material**   有机材料
raw **material**   原材料

We test a wide range of raw **materials**.
我们测试各种各样的原材料。
Material testing helps avoid expensive repair or renovation work.
材料试验有助于避免昂贵的维修或翻新工作。

Day 53

test
★★★★★

/test/

n.&v. 测试，试验

test result 测试结果
hardness test 硬度测试
laboratory test 实验室测试
nondestructive testing 无损检测

We offer comprehensive material **testing** solutions that include field examinations, laboratory **tests** and special inspections.
我们提供全面的材料试验方案，包括现场检查、实验室测试及专检。

The generator will be **tested** for one hour each month.
每月将测试发电机一个小时。

Day 54

component ★★★★★ /kəmˈpoʊnənt/

n. （机器、系统等的）零件，部件

basic **component**	基本组件
electrical **component**	电器件
hydraulic **component**	液压元件
mechanical **component**	机械零件
system **component**	系统部件

We all understand the basic **components** and functions of a centrifugal pump.
我们都了解离心泵的基本组件和功能。

This test proves that all the **components** are working correctly.
该测试证明各个部件运转正常。

This is a factory supplying **components** for the car industry.
这是一家为汽车行业提供零部件的工厂。

Day 55

environmental

/ɪnˌvaɪrən'mentl/
adj. 自然环境的

★★★★★

environmental engineering   环境工程
environmental impact   环境影响
environmental legislation   环境立法

Many of the negative **environmental** impacts of hydroelectric power come from the associated dams.
水力发电对环境产生的许多不利影响都是由与其相关的水坝造成的。

Environmental engineers study the effects of technological advances on the environment to address local and worldwide **environmental** issues.
环境工程师研究技术发展对环境的影响，以解决当地和世界范围内的环境问题。

Day 56

stream
★★★★★

/striːm/

n. 小河，小溪，水流

stream channel	河道
stream erosion	河流侵蚀
stream flow	河流流量

High **stream** density usually means quick surface run-off and flash floods.
河网密度大通常意味着快速的地表径流和骤发的洪水。

The treated water may be discharged into a **stream**, river, bay, lagoon or wetland.
经过处理的水可以排入小溪、河流、海湾、潟湖或湿地。

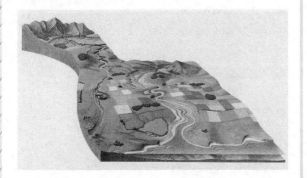

Day 57

fluid
★★★★★

/ˈfluːɪd/
n. 流体，液体

fluid dynamics	流体动力学
fluid mechanics	流体力学
hydraulic **fluid**	液压油
pressurized **fluid**	承压流体

This **fluid** consists primarily of petroleum products.
这种液体主要由石油产品构成。

Hydraulic equipment uses pressurized **fluid** to perform a multitude of machining operation.
液压设备使用承压流体来进行多种加工操作。

Note
液体是物质的一种形态，增温或减压一般能使液体汽化，成为气体（gas）。

Day 58

elevation
★★★★★

/ˌelə'veɪʃən/

***n.** 高程，海拔，立面图*

absolute **elevation**	绝对高程
digital **elevation** model (DEM)	数字高程模型
ground **elevation**	地面高程

DEM can precisely express the ground **elevation**, which makes it play a huge role in every field.
数字高程模型（DEM）能够精确呈现地面高程，这使它在各个领域都发挥巨大作用。

The bed **elevation** is approximately matched with the sea level.
该河床的高程大致与海平面相当。

We're probably at an **elevation** of about 3,900 m above sea level.
我们可能位于海拔大约 3,900 米的高度。

Note
高程指由高程基准面起算的地面点高度。

Day 59

slope
★★★★★

/sloʊp/
n. 斜坡，坡度
v. 倾斜

bottom **slope**	底坡
critical **slope**	临界坡度
longitudinal **slope**	纵向坡度
side **slope**	边坡
steep **slope**	陡坡

You can assume that the bottom of the canal will **slope** downstream.
可以假定运河的底部将向下游倾斜。

Critical **slope** angle exists not only on the **slope** toward the equator but on the **slope** entirely shaded by the sun.
临界坡度角不仅存在于向阳坡，也存在于背阴坡。

Day 60

pressure
★★★★★

/ˈpreʃər/
n. 压力，压强

pressure gauge	压力计
atmospheric **pressure**	气压
hydraulic **pressure**	液压
negative **pressure**	负压

The company is the leading manufacturer of high **pressure** pipes.
该公司是高压管道的主要生产厂商。

Hydraulic transmission takes a liquid as the working medium and uses the **pressure** energy of the liquid to transmit power.
液压传动以液体为工作介质，利用液体的压力能来传递动力。

The gas containers burst at high **pressure**.
这些瓦斯容器遇到高压就会爆炸。

Day 61

basin

★★★★★

/ˈbeɪsən/

n. 盆地，流域

artesian **basin**	自流盆地
detention **basin**	滞洪区
harbor **basin**	港池
Junggar **Basin**	准噶尔盆地
sedimentation **basin**	沉淀池，沉沙池

The Amazon **Basin** is a huge tropical rainforest area in South America that contains the Amazon River and its tributaries.
亚马孙河流域是位于南美洲的巨大热带雨林地区，包含亚马孙河及其支流。

Day 62

rainfall
★★★★★

/ˈreɪnfɔːl/
n. 降雨，降雨量

rainfall duration	降雨历时
rainfall intensity	降雨强度
rainfall run-off	降雨径流
storm **rainfall**	暴雨

Total run-off of a storm is related to the **rainfall** duration and intensity.
一场暴雨的总径流量与降雨历时及强度有关。

When the seasonal **rainfall** is less than the minimum requirement for the satisfactory growth of crops, the irrigation system is essential.
当季降雨量少于作物正常生长所需的最低要求时，灌溉系统就必不可少了。

Day 63

load
★★★★★

/ləʊd/
n. 负荷，荷载

load rejection	甩负荷
electronic **load** controller (ELC)	电子负荷调节器
full **load**	满载，满负荷
suspended **load**	悬移质

A **load** management system is an enhanced version of the electronic **load** controller.
负荷管理系统是电子负荷调节器的增强版。

Sediment is usually divided into two classes: suspended **load** and bed **load**.
泥沙通常分为两类：悬移质和推移质。

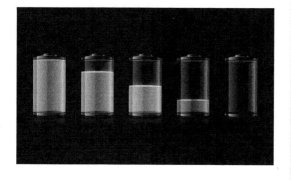

Day 64

filter
★★★★★

/ˈfɪltər/
n. 过滤器
v. 过滤

filter medium	过滤介质
filter press	压滤机
sand **filter**	砂滤器，砂滤池
trickling **filter**	滴滤池

A slow sand **filter** must be operated and maintained properly to make it effective.
必须正确操作和维护慢滤池，从而使其高效运转。

The **filter** bed can be easily cleaned using the **filtered** water.
用过滤后的水可以轻松清洗滤床。

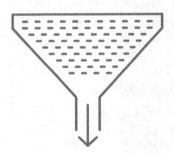

Day 65

chemical

★★★★★

/ˈkemɪkəl/

adj. 化学的

chemical analysis 化学分析
chemical compound 化合物
chemical oxygen demand (COD) 化学需氧量
chemical treatment 化学处理

Chemical treatment may be required to reduce excessive levels of iron, manganese, chalk, and organic matter.
为了降低铁、锰、白垩和有机物质含量，使其恢复到正常水平，可能需要进行化学处理。

Numerous physical, **chemical**, and biological factors affect water quality in ponds, lakes, streams, rivers, oceans, and groundwater.
许多物理、化学和生物因素都会影响池塘、湖泊、溪流、河流、海洋和地下水的水质。

Day 66

depth

★★★★★

/depθ/

n. 深度

depth of flow　　　　　　　水流深度
critical **depth**　　　　　临界深度，临界水深
frost **depth**　　　　　　　　冰冻深度

This irrigation method diverts water to the farmland through small channels, which floods the area up to the required **depth**.
这种灌溉方法通过小渠道将水引到农田里，使该地区达到所需的水深。

Note
临界水深指流量一定而其断面单位能量最小时的水深。

Day 67

crop
★★★★★

/krɑːp/

n. 庄稼，作物

crop growth　　　　　　作物生长
crop rotation　　　　　　庄稼轮作
crop yield　　　　　　　作物产量

Crop production in semi-arid areas faces many risks, including floods.
半干旱地区的作物生产面临包括洪水在内的许多风险。
Good **crop** management reduces water and wind erosion of soil and improves soil fertility.
良好的作物管理可以减少土壤的水蚀和风蚀，提高土壤肥力。

Day 68

source
★★★★★

/sɔːrs/

n. 来源，源头

energy **source**	能量来源
non-point **source** pollution	非点源污染，面源污染
pollution **source**	污染源
power **source**	电源
water **source**	水源

Non-point **source** pollution is water pollution caused by run-off that is not from a single site, such as a wastewater treatment plant or industrial discharge pipe.
非点源污染是指径流造成的水污染，这些径流并非来自废水处理厂或工业排放管道之类的单一地点。

Display the information of the water **source** area and water plant in the form of maps.
用地图显示水源地和水厂信息。

Day 69

level

/ˈlevəl/

n. 水平，水平面

ground **level**	地平面
lake **level**	湖平面
sea **level**	海平面
water **level** monitoring	水位监测

The lake **level** will be dropped to the appropriate **level** before the spring run-off.
湖平面在春季径流前会降到适当的水平。

A number of roads are impacted by high water **levels**.
一些道路受到高水位的影响。

Day 70

hydrolojic
★★★★★

/ˌhaɪdrəˈlɑːdʒɪk/
adj. 水文的

hydrologic analysis 水文分析
hydrologic cycle 水文循环，水循环
hydrologic forecast 水文预报

Designated references are to be used for **hydrologic** analysis of soil and water conservation practices.
指定的参考资料将用于水土保持实践中的水文分析。
The main phases of the **hydrologic** cycle are precipitation, run-off, evaporation, etc.
水循环的主要环节为降水、径流和蒸发等。

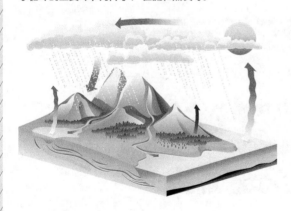

Day 71

field
★★★★★

/fi:ld/

n. 田间，实地，领域

field book   野外工作簿，外业记录簿
field drain   田间排水沟
field testing   田间试验

His theories have not yet been tested in the **field**.
他的理论尚未经过实地检验。

The team is supported by leading experts in the **field** of water management.
这个团队得到水管理领域顶尖专家的支持。

Day 72

condition
★★★★★

/kənˈdɪʃən/

n. 状态，条件

operating **conditions**	操作条件，运行状况
soil **conditions**	土壤条件
weather **conditions**	天气条件，天气状况

Proper management can improve soil **conditions**.
正确的管理可改善土壤状况。

Condition assessment is based on the data obtained.
状态评估的开展基于获取的数据。

Poor working **conditions** lead to demoralized and unproductive employees.
恶劣的工作条件使员工士气低落，生产力降低。

Day 73

tank

★★★★★

/tæŋk/

n. （储存液体或气体的）槽，罐，池等

sedimentation **tank**	沉淀池
septic **tank**	化粪池
storage **tank**	贮藏罐，储槽
surge **tank**	缓冲罐，调压井

On the other hand, the surge **tank** provides excess water.
另一方面，调压井提供多余水量。

If a steel chlorine **tank** is used, it must be painted and checked for corrosion every year.
如果使用钢制氯气罐，则必须每年涂漆并检查是否有腐蚀。

station
★★★★★

/'steɪʃən/

n. 站，台

monitoring **station**   监测站
power **station**   发电站
weather **station**   气象站

The team successfully configured the monitoring **station** over a period of five days.
团队在五天的时间里成功配置了监测站。

A PLC is typically used as the "brain" of a hydropower **station** control platform.
可编程控制器（PLC）通常被用作水电站控制平台的"大脑"。

There are many hydropower **stations** built in the upper reaches of the Yellow River to use the water resources there.
黄河上游修建了许多水电站来利用上游的水资源。

Day 75

site

/saɪt/

n. 场地，工地，现场

★★★★★

site installation	现场安装
site survey	现场勘察
construction **site**	施工工地，施工现场

There follows the second phase of the investigations of the dam **site**.
接下来是对坝址的第二阶段调查。

Over-chutes can be open channel structures (flumes) or pipes, depending on **site** conditions.
根据现场情况，跨渠槽可采用明渠结构（渡槽）或管道。

Day 76

operation
★★★★★

/ˌɑːpəˈreɪʃn/
n. 操作，运行

operation and maintenance 运行和维护
equipment **operation** 设备运行
proper **operation** 正确操作，正常运行

During **operation**, the sand filter is covered with a water layer.
在操作过程中，砂滤池上覆盖着水层。

Water resources assessment is essential to the planning, design, construction, **operation** and maintenance of reliable water systems.
水资源评估对于可靠的水系统的规划、设计、建造、运行和维护至关重要。

Day 77

install
★★★★★

/ɪnˈstɔːl/
v. 安装

installed capacity	装机容量
install sth. in	把……安装在
properly **installed**	已正确安装

China's first intelligent solar-tidal photovoltaic power plant, located in Zhejiang Province, has an **installed** capacity of 100 MW.
中国第一个潮光互补智能光伏电站，位于浙江省，装机容量为 100 兆瓦。

The process of **installing** subsurface drains requires digging deep ditches and **installing** pipes underground.
安装地下排水管道时，需要挖深沟并在地下安装管道。

Day 78

land

★★★★★

/lænd/

n. 土地，地面

land cover	土地覆盖，土地覆被
land ownership	土地所有权
land use	土地利用，土地使用
agricultural **land**	农用地
arable **land**	可耕地

The effect of **land** slope on water volume is usually minor.
地面坡度对水量的影响通常较小。

The **land** area of the world will continue to increase by improving irrigation and drainage.
通过改善灌溉和排水情况，世界土地面积将继续增加。

Day 79

channel

★★★★★

/'tʃænl/

n. 水渠，槽，航道

approach **channel**	引航道
artificial **channel**	人工航道
drainage **channel**	排水渠
open **channel**	明渠

Civil engineers are primarily concerned with open **channel** flow.
土木工程师主要关注明渠水流。

The water flow at any point in a **channel** has two types of energy: potential and kinetic.
水渠中任意一处的水流都具有两种能量，即势能和动能。

Day 80

provide
★★★★★

/prəˈvaɪd/

v. 提供

provide sb. with sth.　　　　向某人提供某物

In most cases, the water flow and fall do not **provide** enough power to drive conventional turbines.
在大多数情况下，水流和落差无法提供足够的动力来驱动常规水轮机。

The main purpose of hydrants is to **provide** water for fire fighting.
消防栓的主要目的是提供消防用水。

Day 81

sewage
★★★★★

/ˈsuːɪdʒ/
n. 污水

sewage treatment	污水处理
domestic **sewage**	生活污水
industrial **sewage**	工业污水

Sewage treatment is the process of removing contaminants from wastewater and household **sewage**.
污水处理指的是从废水和生活污水中去除污染物的过程。

Storm **sewage** means water that is discharged from the surface as a result of rainfall.
暴雨污水是指由于降雨而从地表排出的水。

Day 82

velocity

★★★★★

/vəˈlɑːsəti/

n. 速度

velocity head 流速水头
average **velocity** 平均速度
pulse **velocity** 脉冲速度
water flow **velocity** 水流速度

More complete maps of the **velocity** field are now available with speed sensors.
现在我们可以使用速度传感器获得更完整的速度场图。

This reduces the water flow **velocity**, providing additional time for infiltration.
这降低了水流速度，为渗透提供了更多的时间。

Day 83

capacity
★★★★★

/kəˈpæsəti/
n. 能力，容量

carrying **capacity**	承载能力
heat **capacity**	热容
water-holding **capacity**	持水能力，持水量

Water transport has the largest carrying **capacity** and is the most suitable for carrying bulky goods over long distances.
水路运输的承载能力最强，最适用于大件货物的长途运输。

China's installed **capacity** of renewable energy power generation reached 1 billion kilowatts by the end of October 2021, and the installed hydropower **capacity** stood at 385 million kilowatts.
截至 2021 年 10 月底，中国可再生能源发电装机容量达到 10 亿千瓦，水电装机容量达到 3.85 亿千瓦。

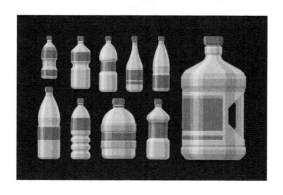

Day 84

drain

/dreɪn/
n. 排水沟，下水管
v. 排水，排出

open **drain** 　　　　　　　　排水明沟
pipe **drain** 　　　　　　　　暗管排水

The overflow of water was caused by a blocked **drain**.
溢水是由排水道堵塞引起的。
As the inflow reduces, the reservoir will begin to **drain** and the water level will drop.
随着入流量的减少，水库将开始排水，水位也将降低。

Day 85

location

/loʊˈkeɪʃən/

★★★★★

n. 地点，位置

geographical **location**	地理位置
horizontal **location**	水平位置
project **location**	项目地点
selected **location**	选址

You may require geotechnical testing in a remote **location**.
你可能需要在某个偏远地点进行岩土工程测试。

Location information shall be appropriately documented.
应该适当记录位置信息。

Day 86

waterway
★★★★★

/ˈwɔːtərweɪ/

n. 航道，水路

artificial **waterway**	人工水道
grass **waterway**	草皮泄水道
inland **waterway**	内河航道
navigable **waterway**	通航水道

The Grand Canal which connects Beijing and Hangzhou in East China's Zhejiang Province, is the longest artificial **waterway** in the world.
大运河连接北京和位于中国东部浙江省的杭州，是世界上最长的人工水道。

They are national leaders in catchment and **waterway** management.
他们是国内流域和航道管理方面的领先者。

Day 87

sediment
★★★★★

/ˈsedəmənt/

n. 沉积物，泥沙

sediment concentration	含沙量
sediment transport	泥沙输移
suspended **sediment**	悬移质泥沙

Relevant data about the content of suspended **sediment** can be calculated using the transport equation of suspended **sediment**.
使用悬移质泥沙输移方程可以计算出悬移质泥沙含量的相关数据。

After obtaining data of bed shear stress, we can evaluate the bed load **sediment** transport rate.
获得床面切应力的数据后，我们就可以估测推移质输沙率。

Day 88

distribution

★★★★★

/ˌdɪstrəˈbjuːʃən/
n. 分配，分布

distribution system 配给系统
rainfall distribution 雨量分布
water distribution 配水

Many kinds of pumps are used in water **distribution** systems.
配水系统中使用许多种泵。
Guangzhou has built the largest "self-healing" electricity distribution grid in China, using 6,242 public feeders in the power **distribution** network.
广州使用配电网中 6,242 条公用馈线，建成了全国最大的自愈配电网。

Day 89

infiltration
★★★★★

/ˌɪnfɪlˈtreɪʃən/

n. 渗透，入渗

infiltration capacity 渗入量
infiltration gallery 渗渠
infiltration rate 入渗率

Different types and locations of soil have different rates of **infiltration**.
土壤类型和位置不同，入渗率不同。
As **infiltration** continues, some of the voids between soil particles are filled with water.
随着渗透的持续进行，土壤颗粒间的某些空隙被水填满了。

Day 90

survey
★★★★★

/ˈsɜːrveɪ/

n. 调查，测量

China Geological **Survey**	中国地质调查局
hydraulic engineering **survey**	水利工程测量
hydrographic **survey**	海道测量，水文测量
topographical **survey**	地形测量

In some cases, this program has eliminated the need for a bridge site **survey**.
在某些情况下，有了这个程序就不需要进行桥渡勘测了。

An international **survey** found that more than half of the citizens considered the climate crisis as the main environmental challenge the world was facing now.
一项国际调查显示，一半以上的民众认为气候危机是当今世界面临的主要环境挑战。

Day 91

natural

★★★★★

/ˈnætʃərəl/
adj. 自然的，天然的

natural disaster　　　　　自然灾害
natural environment　　　　自然环境
natural resources　　　　　自然资源

Executive orders have emphasized the need to conserve **natural** resources and improve the environment.
行政命令强调了保护自然资源、改善环境的必要性。
Landslides are mostly caused by **natural** factors such as heavy rain and steep slopes.
山体滑坡主要是由自然因素（如暴雨、陡坡）造成的。

Day 92

downstream

★★★★★

/ˌdaʊnˈstriːm/
adv. 在下游
adj. 在下游的

downstream industry	下游产业
drift **downstream**	顺流漂下

Downstream river conditions can have a large effect upon flow and flow depth.
下游的河流条件对流量和水流深度影响很大。

This allows the plant to quickly restore the flow **downstream** even if the units could not resume operation.
这样一来，即使机组无法恢复运行，水电厂也能迅速恢复下游水流。

Note
前缀 down- 与名词、动词结合，构成新词，描述从高处向低处进行或位于地势较低处的事物，如 downstairs、downriver 和 downstage。

factor

★★★★★

/ˈfæktər/

n. 因素，要素，系数

environmental **factor**	环境因素
load **factor**	负载系数
plant **factor**	设备利用率
power **factor**	功率因数

If the turbine is only used for domestic lighting in the evenings, the plant **factor** will be very low.
如果涡轮机只用于晚上家用照明，设备利用率就会很低。

The frost resistance of the concrete structure is a very important **factor** in the construction of buildings.
混凝土结构的抗冻性是建筑施工中一个非常重要的因素。

Day 94

hydrograph
★★★★★

/ˈhaɪdrəˌɡræf/
n. 水文过程线

flood **hydrograph**	洪水过程线
synthetic **hydrograph**	综合水文过程线
unit **hydrograph**	单位过程线，单位线

A dam breach **hydrograph** can be natural or synthetic.
溃坝过程线可以是自然的或合成的。

Precipitation plays a key role in determining peak discharges and run-off, as well as in **hydrograph** development.
降水量对于洪峰流量和径流的确定以及水文过程线的绘制，均起着关键作用。

Day 95

penstock
★★★★★

/ˈpenˌstɑːk/

n. 压力管道

steel **penstock** 压力钢管

In a **penstock**, the velocity of water flow changes suddenly due to some external reasons, forming the phenomenon of water hammer.
在压力管道中，由于某些外界原因，水的流速突然发生变化，从而引起水击。

Penstocks can be used to transmit gases and liquids.
压力管道可用于输送气体和液体。

measure
★★★★★

/ˈmeʒər/
n. 办法，措施
v. 测量

conservation **measure**	保护措施
control **measure**	控制措施
mitigation **measure**	减缓措施
remedial **measure**	补救措施

The improvement in productivity is due to the increased adoption of soil conservation **measures** by the farmers.
生产力的提高归功于农民逐渐采取了土壤保护措施。
A pressure indicator is used to **measure** hydraulic pressure at any point.
压力指示器用于测量任意一点的液压。

Day 97

electrical
★★★★★

/ɪˈlektrɪkəl/
adj. 有关电的，电气的

electrical fault　　　　　　电力故障
electrical system　　　　　电气系统

All **electrical** wiring shall be done in accordance with requirements from local authorities.
所有电气布线应按照当地要求进行。

The **electrical** substation is an integral part of a typical hydroelectric station.
变电站是典型水电站不可缺少的一部分。

Day 98

moisture
★★★★★

/ˈmɔɪstʃər/
n. 水分，湿气

moisture index 湿润指数
soil **moisture** monitoring 墒情监测

Monitoring soil **moisture** conditions provides important information for the protection of local and regional water resources.
监测墒情可为保护地方和区域水资源提供重要信息。
It is hard to manually track soil temperature and **moisture** levels.
很难人工跟踪土壤温度和湿度水平。

Day 99

electricity

★★★★★

/ɪˌlekˈtrɪsəti/
n. 电

electricity consumption	电力消耗，用电
electricity meter	电表
electricity supply	电力供应

They usually release the stored water to generate **electricity** during peak **electricity** demand periods.
他们通常会在电力需求高峰期释放储存的水来发电。

Hydropower plants can supply large amounts of **electricity,** and they are relatively easy to adjust to demands.
水电厂可以提供大量的电，并且可以根据需求相对容易地进行调整。

Day 100

第二部分
四星词汇

pollution
★★★★

/pəˈluːʃən/
n. 污染

air **pollution**	空气污染
fecal **pollution**	粪便污染
noise **pollution**	噪声污染
water **pollution**	水污染

Groundwater **pollution** most often results from improper disposal of wastes on land.
地下水污染最常见的形成原因是地面废物处理不当。

As a large amount of untreated industrial waste being dumped straight into water systems, **pollution** management is a must.
因为有大量未经处理的工业废物直接排入水系，所以污染管理必不可少。

Day 101

mechanical
★★★★

/mɪˈkænɪkəl/
adj. 机械的，力学的

mechanical energy　　　　　机械能
mechanical governor　　　　机械式调速器
mechanical property　　　　 力学性能

Replacing poorly performing **mechanical** components is very costly.
更换性能不佳的机械部件开销很大。

Note
-al 与名词结合，构成形容词，描述事物与原名词表示的事物有关，如 accident—accidental、profession—professional。

Day 102

catchment

★★★★

/ˈkætʃmənt/

n. 集水区，流域

catchment area　　　　集水面积，流域面积

Run-off in the **catchment** is influenced by hydrologic factors such as topography.
集水区的径流受地形等水文因素的影响。

All electricity is generated by solar panels, while **catchment** basins store fresh rainwater for all the retreat's needs.
所有的电都是太阳能电池板产生的，而集水区储存着新鲜的雨水，可以满足度假区的所有需求。

Day 103

requirement
★★★★

/rɪˈkwaɪrəmənt/

n. 要求，需求

basic **requirement**	基本要求
regulatory **requirement**	监管要求
specific **requirement**	具体要求，特殊要求

Safe drinking water and water for domestic use are basic **requirements** of households and communities.

安全饮用水和家庭用水是家庭和社区的基本需求。

Note

词尾 -ment 可以加在许多动词之后构成名词，如 develop—development、pay—payment、amaze—amazement。

protection
★★★★

/prəˈtekʃən/
n. 保护，防护

bank **protection**	护岸
environmental **protection**	环境保护
fire **protection**	消防，防火
flood **protection**	防洪

As for flood **protection**, umbrellas and sandbags sometimes are not efficient, as they fail to provide adequate shelter from torrential downpours and floodwater.
在防洪方面，雨伞和沙袋有时并没有用，因为在暴雨和洪水来临时，它们不能提供足够的庇护。

An overcurrent **protection** device is a piece of electrical equipment used to protect service, feeder, and branch circuits and equipment from excess current by interrupting the flow of current.
过电流保护装置是一种电气设备，通过中断电流来保护业务通信电路、馈线电路、分支电路和设备免受过大电流的影响。

Day 105

filtration
★★★★

/fɪlˈtreɪʃən/
n. 过滤

direct **filtration** 直接过滤
membrane **filtration** 膜过滤
sand **filtration** 砂滤

Filtration gathers together impurities that float on water and boosts the effectiveness of disinfection.
过滤可以聚集浮在水面上的杂质，有效消毒。

Day 106

lake
★★★★

/leɪk/

n. 湖，湖泊

artificial **lake**	人工湖
West **Lake**	西湖

Lake water level conditions indicate a higher risk of major spring floods in those areas.
湖泊水位状况表明，那些地区发生重大春季洪水的风险更高。

Note
世界最大湖泊为里海（Caspian Sea）。
世界面积最大淡水湖为苏必利尔湖（Superior Lake）。
中国最大内陆咸水湖为青海湖（Qinghai Lake）。
中国最大淡水湖为鄱阳湖（Poyang Lake）。

Day 107

available
★★★★

/əˈveɪləbəl/

adj. 可获得的，现有的，可用的

available data　　　　现有数据
readily **available**　　　现成的，便捷可得的

The choice of turbines depends mainly on the pressure heads **available**.
涡轮机的选择主要取决于有哪些可用的压力水头。
Further building can continue when money is made **available**.
有了资金就可以继续进一步的施工。

Day 108

include
★★★★

/ɪnˈkluːd/

v. 包括

Municipal water supply systems **include** facilities for storage, transmission, treatment, and distribution.
市政供水系统包括水的储存、输送、处理和分配等设施。
Specialized inspections **include** earthquake and check flood inspections.
专业检查包括防震检查和校核洪水检查。
The curriculum **includes** courses in hydraulics.
课程内容包括水力学。

Day 109

installation

★★★★

/ˌɪnstəˈleɪʃən/
n. 安装，装置

electrical **installation**	电气安装
equipment **installation**	设备安装

Electrical **installation** in the powerhouse needs good workmanship.
发电厂的电气安装需要精湛的技艺。

Day 110

peak ★★★★

/piːk/
n. 峰值，顶点
adj. 最好（高）的

peak discharge　　　　　　　　洪峰流量
peak load　　　　　　　　　　　高峰负荷

The **peak** demand for electricity in residential areas usually occur in the morning and early evening hours.
早晨和傍晚通常是居民区的用电需求高峰时段。

Note
洪水总量（flood volume）和洪峰流量（peak discharge）是衡量洪水量级大小的重要指标。

Day 111

weir ★★★★

/wɪr/
n. 堰，拦河坝

weir flow　　　　　　　　　　　堰流
Dujiang **Weirs**　　　　　　　　　都江堰
rectangular **weir**　　　　　　　　矩形堰

The **weir** has two enormous arched gates to control water.
该堰设有两个大的拱形闸门，用以控制水流。

Day 112

operate
★★★★ /ˈɑːpəreɪt/
v. 操作，运转，运作

operate automatically 自动操作

Reverse osmosis (RO) **operates** at relatively high pressures and can be used to remove dissolved inorganic compounds from water.
反渗透（RO）运作于相对高压的地方，可用于从水中分离出溶解性无机化合物。

Generally, two pumps will be **operated** and a third pump is there for backup.
通常会有两个泵运转，第三个泵作为备用。

Day 113

process
★★★★ /ˈprɑːses/
n. 过程，工艺

aerobic **process** 需氧过程
biological **process** 生物过程

Desalination **processes** are used in areas where freshwater supplies are not readily available.
脱盐工艺在没有现成淡水供应的地区使用。

Day 114

normal
★★★★

/ˈnɔːməl/
adj. 正常的，正规的，标准的
n. 正常

normal distribution	正态分布
above **normal**	超常，高于正常水平
below **normal**	低于正常水平

The river is forecasted to have water flow at the near **normal** level to slightly above **normal** until the spring run-off.
在春季径流到来之前，这条河的流量预计将保持在接近正常到略高于正常的水平。

Day 115

quantity
★★★★

/ˈkwɑːntəti/
n. 量，数量

a large **quantity** of	大量的
adequate **quantity**	足够数量
vector **quantity**	矢量，向量

A large **quantity** of fresh water is being consumed daily by food processing.
食品加工每天消耗大量淡水。

Day 116

measurement
★★★★
/ˈmeʒərmənt/
n. 测量，量度，计量

quantitative **measurement**   定量测量

Relative Humidity **Measurement** Systems offer a complete profile of moisture content throughout the entire slab.
相对湿度测量系统可提供整个板坯中水分含量的完整信息。

Day 117

minimum
★★★★
/ˈmɪnəməm/
adj. 最小的，最少的

minimum flow   最小流量，最低流量
minimum requirement   最低要求
minimum tillage   少耕法

Contour strip cropping combined with crop rotation and **minimum** tillage is an effective method of soil and water conservation.
将等高带状种植与轮作及少耕法结合是有效的水土保持方法。

Day 118

necessary
★★★★
/ˈnesəseri/
adj. 必要的，必需的

necessary for 对……有必要
if necessary 如有必要

If **necessary**, consult a professional before installing a system.
如有必要，在安装系统之前咨询专业人士。

Day 119

bacteria
★★★★
/bækˈtɪriə/
n. 细菌

aerobic **bacteria** 需氧细菌，好氧细菌
coliform **bacteria** 大肠菌群

Chlorine is added to kill **bacteria**.
添加了氯来杀灭细菌。

Day 120

well ★★★★ /wel/
n. 井，井道

well casing	井筒
artesian **well**	自流井
wet **well**	湿井

If there is any source of chemical pollution near the construction area of the **well**, the local public health authority should be contacted.
如果井的施工区域附近有任何化学污染源，应联系当地公共卫生部门。

Day 121

voltage ★★★★ /ˈvoʊltɪdʒ/
n. 电压，伏（特）

voltage regulator	电压调节器，稳压器
high **voltage** (HV)	高压
low **voltage** (LV)	低压
medium **voltage** (MV)	中压
rated **voltage**	额定电压

For equipment with small capacities, **voltage** protection is optional.
对于小容量的设备来说，电压保护装置并非必选。

Day 122

四星词汇

facility
★★★★ /fəˈsɪləti/
n. 设施,设备

hydropower **facilities** 水电设施
industrial **facilities** 工业设施
storage **facilities** 贮存设备

Assessing existing equipment is a key aspect of upgrading **facilities**.
评估现有设备是设备升级的一个关键环节。

Day 123

ensure
★★★★ /ɪnˈʃʊr/
v. 确保,保证

Monitor equipment operation and performance and make necessary adjustments to **ensure** optimal performance.
监控设备的运行和性能,并进行必要调整,以确保设备保持最佳性能。

Day 124

harbor
★★★★

/ˈhɑːrbər/
n. 港湾，港口，码头

fishery **harbor** 　　　　　　　　　渔港
natural **harbor** 　　　　　　　　　天然港

The **harbor** master should ensure appropriate water treatment within the fishery **harbor**.
港务局长应确保渔港内的水处理是恰当的。

Day 125

upstream
★★★★

/ˌʌpˈstriːm/
adv. 在上游
adj. 上游的

upstream face 　　　　　　　　　上游面

Water will flow in canals as long as the water level is higher at the end **upstream** than the end downstream.
只要上游端水位高于下游端水位，水就会在运河中流动。

Day 126

contamination ★★★★
/kənˌtæməˈneɪʃən/
n. 污染，污染物

chemical **contamination**   化学污染
lead **contamination**   铅污染
microbial **contamination**   微生物污染

In industrialized countries, the public concern has been shifted to the chronic effects on health related to chemical **contamination**.
在工业化国家，人们已经开始关注化学污染引起的慢性疾病。

Day 127

rate ★★★★
/reɪt/
n. 率，速率

flow **rate**   流量
infiltration **rate**   入渗率

When the **rate** of rainfall exceeds the infiltration **rate**, the excess water will start flowing over the soil surface.
当降雨率超过入渗率时，多余的水就开始在土壤表面流动。

Day 128

assessment ★★★★

/əˈsesmənt/
n. 评价，评估

risk **assessment** 风险评估，危险性评估
water resources **assessment** 水资源评价

In order to conduct a comprehensive water resources **assessment**, you need data from various experts from different backgrounds.
为了开展全面的水资源评价，你需要来自不同背景的多位专家收集的数据。

Day 129

vertical ★★★★

/ˈvɜːrtɪkəl/
adj. 垂直的，纵向的

vertical axis 立轴
vertical distance 垂直距离
vertical pump 立式泵

Safety measures usually include **vertical** and horizontal channel stabilization.
安全措施通常包括保持垂直和水平通道稳定。

Day 130

energy
★★★★

/ˈenərdʒi/
n. 能，能源

kinetic **energy**	动能
potential **energy**	势能
renewable **energy**	可再生能源

Potential **energy** of the water is converted into kinetic **energy** as it flows down the penstock due to gravity.
水在重力作用下流经压力管道时，势能转化为动能。

Day 131

repair
★★★★

/rɪˈper/
n. 修理，维修

repair service	维修服务
maintenance and **repair**	维护和修理

Carry out equipment maintenance and **repair** as necessary.
必要时，对设备进行维护和修理。

Day 132

transport
★★★★

/ˈtrænspɔːrt/
n. 运输，输送，传递

inland water **transport**	内河运输
mode of **transport**	运输方式
ocean **transport**	远洋运输

Water **transport** is the cheapest and the oldest mode of **transport**.
水运是最便宜、最古老的运输方式。

Day 133

solid
★★★★

/ˈsɑːlɪd/
n. 固体
adj. 固体的，坚硬的

settleable **solid**	可沉降固体物
suspended **solid**	悬浮物，悬浮固体
total dissolved **solid** (TDS)	总溶解固体

Absorption is the capacity of a **solid** particle to attract molecules to its surface.
吸附是指固体颗粒将分子吸引到其表面的能力。

Day 134

inspection ★★★★
/ɪnˈspekʃən/
n. 检验，检查

regular **inspection**	定期检查
safety **inspection**	安全检查
visual **inspection**	目视检查，外观检验

Regular **inspection** of runners of turbines should be carried out and records should be invariably maintained.
应当定期检查水轮机的转轮，并坚持作好记录。

Day 135

organic ★★★★
/ɔːrˈɡænɪk/
adj. 有机的

organic compound	有机化合物
organic matter	有机质，有机物

As **organic** matter decays, it uses up oxygen.
有机物腐烂时会耗尽氧气。

Sludge is often inadvertently contaminated with many toxic **organic** and inorganic compounds.
沉积物常常在不经意间被很多有毒的有机和无机化合物污染。

Day 136

information ★★★★
/ˌɪnfərˈmeɪʃən/
n. 信息，资料，情报

available **information** 可用信息
geographical **information** system (GIS) 地理信息系统
hydrological **information** 水文情报

Hydrological **information** can be obtained from the meteorology or irrigation department.
可以从气象或灌溉部门获取水文情报。

Day 137

industrial ★★★★
/ɪnˈdʌstriəl/
adj. 工业的

industrial city 工业城市
industrial wastewater 工业废水

The **industrial** revolution grew our capacity to put up buildings and infrastructure exponentially.
工业革命飞速提高了我们建造建筑和基础设施的能力。

Day 138

relay
★★★★
/ˈriːleɪ/
n. 继电器

polarized **relay** 极化继电器
power **relay** 功率继电器
protective **relay** 保护继电器

High currents or voltages of the electric power system cannot be directly fed to **relays** and meters.
电力系统的高电流或高电压不能直接馈入继电器和电表。

Day 139

inlet
★★★★
/ˈɪnlet/
n. 入口，进水口

drop **inlet** 落底式进水口
submerged **inlet** 潜没式进水口

Raw water slowly enters the filter through an **inlet**.
原水通过进水口缓慢流入过滤器。
Close the **inlet** valve and the guide vane completely.
将进水阀和导叶完全关闭。

Day 140

contaminant ★★★★
/kən'tæmɪnənt/
n. 污染物，杂质

contaminant transport　　污染物运移
organic **contaminant**　　有机污染物

Modern testing methods allow the detection of air **contaminants** in extremely low concentrations.
现代检测方法可以检测出极低浓度的空气污染物。
Self-purification is the ability of rivers to purify itself of **contaminants** by natural processes.
自净是指河流通过自然过程净化自身杂质的能力。

Day 141

spillway ★★★★
/'spɪlweɪ/
n. 溢洪道

drop **spillway**　　跌水式溢洪道
ogee **spillway**　　反弧面溢洪道
side channel **spillway**　　侧槽式溢洪道

Spillways may be set up at one side or both sides of the dam.
溢洪道可能会设在大坝单侧或两侧。

Day 142

disinfection
★★★★

/ˌdɪsənˈfekʃən/
n. 消毒，灭菌

chemical disinfection 化学消毒

Conventional treatment methods of water involves clarification, **disinfection**, etc.
传统的水处理方法包括澄清和消毒等。

This test is one of the several tests that are used to measure the **disinfection** efficiency of pools and spas.
这项测试是衡量泳池和水疗中心消毒效率的几项测试之一。

Day 143

forecast
★★★★

/ˈfɔːrkæst/
n.&v. 预报，预测

flood forecast 洪水预报

Updated **forecast** information will be posted on flood sheets regularly.
更新的预报信息将定期发布在洪水报表上。

The Hydrologic **Forecasting** Center will continue to monitor watershed conditions closely.
水文预报中心将继续密切监控流域状况。

Day 144

agricultural
★★★★
/ˌæɡrɪˈkʌltʃərəl/
adj. 农业的，农学的

agricultural drainage　　　　农业排水
agricultural production　　　农业生产
agricultural purpose　　　　 农业用途

Sprinklers have been increasingly used for irrigation in **agricultural** land.
喷灌器已越来越多地用于农田灌溉。

Day 145

remove
★★★★
/rɪˈmuːv/
v. 去除，消除，排除

remove impurities　　　　　　去除杂质

Water erosion can impact water quality and kill fish by **removing** sediments and nutrients.
水蚀会因其去除了泥沙和营养物质而影响水质并导致鱼类死亡。
The filter will **remove** most impurities found in water.
过滤器会滤掉水中的大部分杂质。

Day 146

application ★★★★

/ˌæplɪˈkeɪʃən/
n. 使用，应用

engineering **application**	工程应用
industrial **application**	工业应用
software **application**	软件应用

Most industrial **applications** use electric motors as prime movers to rotate hydraulic pumps.
大多数的工业应用都将电机作为转动液压泵的原动机来使用。

Day 147

turbidity ★★★★

/tɜːrˈbɪdəti/
n. 浊度，混浊

low/high **turbidity**	低/高浊度
nephelometric **turbidity** unit	散射浊度单位

Turbidity in water is caused by suspended solids and colloidal matter.
水的混浊是由悬浮物和胶体物质造成的。
Water with high **turbidity** causes high filtration costs.
浊度高的水过滤费用昂贵。

Day 148

cause ★★★★

/kɔːz/
v. 引起，使发生
n. 原因

cause damage	造成损害
cause diseases	致病
cause problems	引起问题
main **cause**	主要原因

Silica inhaling may **cause** lung problems.
吸入二氧化硅可能会导致肺部疾病。
Surface run-off is the principal **cause** of soil erosion.
地表径流是土壤侵蚀的主要原因。

Day 149

gate ★★★★

/geɪt/
n. 闸门，大门

control **gate**	控制闸门
lock **gate**	船闸闸门
sluice **gate**	泄水闸门
turbine **gate**	水轮机闸门

The amount of water which is to be released from the penstock can be controlled by a control **gate**.
可以通过控制闸门来控制压力管道中释放的水量。

Day 150

depend
★★★★
/dɪˈpend/
v. 依赖，依靠，取决于

depend on/upon　　　依赖，依靠，取决于

The transport system will **depend** on gravity flow as far as possible.
运输系统将尽可能依赖重力流。

Optimum power generation and the performance of a scheme **depend** upon the design and selection of electrical systems and components.
最佳发电量和方案的效益是由对电气系统和电气部件的设计和选择决定的。

Day 151

sewer
★★★★
/ˈsuːər/
n. 下水道，污水管，阴沟

sanitary **sewer**　　　污水管，生活污水管道

Water and **sewer** bills can be paid by credit card or by bank draft.
水费和下水道费可以通过信用卡或银行汇票支付。

If you have any questions about **sewer** charges, please contact our office.
如果您对下水道收费有任何疑问，请与我们办公室联系。

Day 152

crane
★★★★
/kreɪn/
n. 起重机，吊车

cable **crane**	缆索起重机
overhead **crane**	高架起重机，房顶吊车
tower **crane**	塔式起重机

A tower **crane** is a fixed **crane** that is used for hoisting heavy materials for tall structures.
塔式起重机是一种固定式起重机，用于吊起高层建筑所用的重型材料。

The cable **crane** system monitors the entire track through surveillance cameras.
该缆索起重机系统通过视频监控整个轨道。

Day 153

transformer
★★★★
/trænsˈfɔːrmər/
n. 变压器

single phase **transformer**	单相变压器
step-down **transformer**	降压变压器
step-up **transformer**	升压变压器

The **transformer** inside the powerhouse converts alternating currents to higher-voltage currents.
发电厂内的变压器将交流电转换成电压更高的电流。

Day 154

riparian
★★★★

/raɪˈperiən/
adj. 河岸的，河岸权的

riparian land   沿岸地
riparian right   河岸权

The **riparian** right has its roots in the English Common Law.
河岸权来源于英国普通法。

Day 155

sand
★★★★

/sænd/
n. 砂土，泥沙

coarse **sand**   粗砂
fine **sand**   细砂
slow **sand** filter   慢滤池

Slow **sand** filters use biological methods to purify raw water and produce potable water.
慢滤池使用生物法来净化原水，生产饮用水。
A double-layer filter consists of a layer of anthracite coal and a layer of fine **sand**.
双层滤料滤池由一层无烟煤和一层细砂组成。

Day 156

storm
★★★★
/stɔːrm/
n. 暴雨，风暴

storm drain　　　　　　　　　　雨水沟
storm surge　　　　　　　　　　风暴潮

This flood barrier is one of the largest moving structures in the world that can protect local people during the **storm**.
这座拦洪坝是世界上最大的活动建筑之一，能在暴风雨时保护当地居民。

Day 157

analysis
★★★★
/əˈnælɪsɪs/
n. 分析

risk **analysis**　　　　　　　　　风险分析
statistical **analysis**　　　　　　统计分析

The statistical **analysis** system includes a historical data query module, a data report statistics module, and a data **analysis** comparison module.
统计分析系统包括历史数据查询模块、数据报表统计模块和数据分析比较模块。

Day 158

aquifer
★★★★

/ˈækwɪfər/
n. 含水层

artesian **aquifer**   自流含水层

Some **aquifers** are actually higher than the surrounding ground surface, which may result in artesian springs or artesian wells.
一些含水层实际上要高出周围地面，这可能会形成自流泉或自流井。

Note
aquifer 的前缀 aqu- 来源于拉丁语中的 aqua，意思是"水"。带有该前缀的词语多与水相关，例如，aquatic 意为"水生的"。

Day 159

obtain
★★★★

/əbˈteɪn/
v. 获得，获取

obtain data   获取数据
obtain information   获取信息

Sheet erosion data may be **obtained** in several different ways.
片蚀数据可以通过几种不同的方式获取。

Day 160

shaft ★★★★
/ʃæft/
n. 竖井，轴

shaft drainage 竖井排水
gate **shaft** 闸门井
vertical **shaft** 竖井，立轴

A generator is set up in the powerhouse and it is mechanically coupled to the turbine **shaft**.
发电机安装在发电站里，通过机械装置连接到水轮机轴上。

Day 161

contour ★★★★
/ˈkɑːntʊr/
n. 等高线

contour farming 等高耕种
contour interval 等高距

The **contour** interval will be shown on the plan in the form of notes.
等高距会在平面图上以注释的形式显示。
Contour farming involves plowing, planting and weeding along the **contour** across the slope.
等高耕种包括沿着斜坡上的等高线进行耕地、种植和除草。

Day 162

solution
★★★★
/sə'luːʃən/
n. 溶液，解决方法

chlorine **solution** 　　　　　　　氯溶液
optimal **solution** 　　　　　　　最优解

The soil **solution** contains various soluble substances.
土壤溶液中含有各种可溶性物质。

Day 163

technical
★★★★
/'teknɪkəl/
adj. 技术的，工艺的

technical assistance 　　　　　技术援助
technical regulation 　　　　　技术规程
technical standard 　　　　　　技术标准
technical support 　　　　　　　技术支持

We've been having some **technical** problems with the new hardware.
我们在这个新硬件上遇到了一些技术问题。
These **technical** regulations are clear and precise.
这些技术规程清晰准确。

Day 164

standard ★★★★

/ˈstændərd/
n. 标准，规范

drinking water **standard**	饮用水标准
industrial **standard**	行业标准
national **standard**	国家标准
water quality **standard**	水质标准

In some countries, water quality **standards** may be stipulated by national legislation.
在有些国家，水质标准可能是由国家法律规定的。

Day 165

profile ★★★★

/ˈprəʊfaɪl/
n. 剖面，断面，轮廓

flow **profile**	流动剖面
longitudinal **profile**	纵剖面
soil **profile**	土壤剖面

Infiltration is the movement of water into the soil **profile** from the soil surface.
入渗指的是水经由土壤表面流向土壤剖面的过程。

Day 166

breaker
★★★★
/ˈbreɪkər/
n. 断路器

circuit **breaker**	断路器
generator **breaker**	发电机断路器

A circuit **breaker** is a mechanical device to open and cut off currents under normal and abnormal conditions.
断路器是一种机械装置，能在正常或异常情况下接通、切断电流。

Day 167

oxygen
★★★★
/ˈɑːksɪdʒən/
n. 氧

biochemical **oxygen** demand (BOD)	生化需氧量
dissolved **oxygen** (DO)	溶解氧
oxygen level	氧含量

DO can measure the amount of **oxygen** in water available to the flora and fauna.
溶解氧可以衡量水中动植物可用的氧气量。
Biochemical **oxygen** demand and dissolved **oxygen** can be used as water quality indicators.
生化需氧量和溶解氧可以作为衡量水质的指标。

Day 168

drip

/drɪp/
n. 水滴
v. 滴水

drip irrigation　　　　　　　　　　滴灌

When we irrigate land with a **drip** irrigation system, water **drips** slowly and evenly into the soil from the dripper.
当我们采用滴灌系统进行灌溉时，水滴会缓慢且均匀地从滴头流向土壤。

Day 169

plan

/plæn/
n. 平面图，计划，方案

plan view　　　　　　　　　　　平面图
project management **plan**　　　项目管理方案
site **plan**　　　　　　　　　　　场地平面图

A solid line representing the location of this cross-section shall be drawn on the site **plan**.
需用实线在场地平面图上标注出横截面的位置。

Day 170

cross-section
★★★★
/ˈkrɒːs ˌsekʃən/
n. 横截面，剖面

cross-section area　　横截面积

Day 171

suitable
★★★★
/ˈsuːtəbəl/
adj. 适宜的，合适的

suitable conditions　　适宜条件
suitable for sb./sth.　　适合某人/某物

Concrete penstocks are **suitable** for low heads.
混凝土压力管道适用于低水头。
Choosing the most **suitable** source for water supply largely depends on local conditions.
选择最合适的供水水源很大程度上依赖于当地的条件。

Day 172

perform
★★★★ /pərˈfɔːrm/
v. 履行，执行

perform an analysis 进行分析
perform well/badly 表现得好/差

This position **performs** related duties as required.
这个岗位要按照规定履行相关职责。
These professionals can **perform** land surveys, design conservation plans and monitor conditions.
这些专业人员可以进行土地调查、制定保护计划并监测状态。

Day 173

machinery
★★★★ /məˈʃiːnəri/
n. 机器，机械

construction **machinery** 工程机械
heavy **machinery** 重型机械
hydraulic **machinery** 水力机械，液压机械

Most heavy construction **machinery** utilizes hydraulic power.
大多数重型施工机械利用水力。

Day 174

maximum ★★★★

/ˈmæksəməm/
adj. 最大值的，最大量的
n. 最大值，最大限度

maximum capacity　　　　　　最大容量
a **maximum** of　　　　　　　　最大值

Be careful not to exceed the **maximum** permissible average velocity.
注意不要超过容许的最高平均速度。
The height of the dam always depends on the **maximum** reservoir capacity.
大坝的高度总是根据水库的最大库容确定的。

Day 175

siphon ★★★★

/ˈsaɪfən/
n. 虹吸，虹吸管

siphon rainfall recorder　　　　虹吸式雨量计

A **siphon** is a bent tube used for getting liquid out of a container; often when holding it, let one end of the tube at a lower level than the end in the container.
虹吸管是用来从容器中吸取液体的弯管，一般手持时，要让一端比容器里的另一端位置更低。

Day 176

subsurface
★★★★
/ˈsʌbˌsɜːrfɪs/
adj. 地面下的

subsurface drainage system　　地下排水系统
subsurface water　　地下水

Water supplies may be classified as surface or **subsurface** water.
给水可分为地表水和地下水。

Note
subsurface 的前缀 sub- 常用在形容词、名词前，表示"下面"，如 subway（地铁）、submarine（潜水艇）。

Day 177

device
★★★★
/dɪˈvaɪs/
n. 装置，设备，仪器

mechanical **device**　　机械装置
protection **device**　　保护装置

The protection **device** shall provide additional protection against electric shock.
保护装置应提供额外的防电击保护。
Connect the pipes from one control **device** to another.
将这些管道从一个控制装置接到另一个控制装置上。

Day 178

particle
★★★★

/ˈpɑːrtɪkəl/
n. 颗粒，粒子

soil **particle**　　　　　　　　土壤颗粒
solid **particle**　　　　　　　　固体颗粒
suspended **particle**　　　　　悬浮颗粒

Vegetation and crop residues protect the soil surface from raindrop impacts and reduce **particle** fragmentation.
植被和作物残茬可以保护土壤表面不受雨滴冲击，减少土壤颗粒破碎。

Day 179

size
★★★★

/saɪz/
n. 大小，尺寸

On occasion, the drops may be of drizzle **size**.
有时，水滴可能如细雨点般大小。
Dams are of different **sizes** and shapes, and are made of various materials.
大坝的尺寸、形状和材料多种多样。

Day 180

waste
★★★★

/weɪst/
n. 废物，浪费

waste disposal — 废物处理
hazardous **waste** — 有害废物
liquid **waste** — 液体废物
solid **waste** — 固体废物

Hazardous medical **waste** could lead to contamination of water sources.
有害医疗废物可能导致水源污染。

Day 181

prevent
★★★★

/prɪˈvent/
v. 预防，防止

prevent erosion — 防止侵蚀
prevent flooding — 预防水灾，防洪

Flood control refers to all methods used to reduce or **prevent** the detrimental effects of floodwater.
防洪是指为减轻或防止洪水带来的不利影响所采取的全部措施。

Day 182

map
★★★★
/mæp/
n. 地图

geological **map**	地质图
soil **map**	土壤图
topographic **map**	地形图

According to the **map** we should turn left.
从地图上看，我们应该左拐。

Day 183

estimate
★★★★
/ˈestəmət/ *n.* 估算，估价
/ˈestɪmeɪt/ *v.* 估计，估算

cost **estimate**   成本估算

Visit the channel and look for high water marks to **estimate** flood flows.
巡视航道并寻找高水位标志，以估算洪水流量。
When developing a cost **estimate**, the estimator should always consider the constructability and the construction schedule.
进行成本估算时，估算人员应始终考虑施工可能性和施工进度。

Day 184

district
/ˈdɪstrɪkt/
★★★★
n. 区域，地区

conservation **district**   保护区
drainage **district**   排水区

The water demand in commercial and industrial **districts** is usually evenly distributed during the working day.
在工作日，商业区和工业区用水需求的分布通常比较均衡。

Day 185

maintain
/meɪnˈteɪn/
★★★★
v. 保持，维持，维修

maintain contact with   与……保持联系

Maintain a close customer relationship with clients on all designated contracts.
与客户就所有指定合同保持密切联系。
Watercourses need a certain amount of the water flow to **maintain** healthy conditions.
水道需要一定量的水流来维持其良好状况。

Day 186

四星词汇 151

intake
★★★★ /ˈɪnteɪk/
n. 进水口，入口

water intake 进水口，取水口

A water **intake** structure is built within the reservoir.
水库内建有进水建筑物。
A multilevel **intake** allows water of best quality to be withdrawn.
分层式进水口可以提取最佳水质的水。

Day 187

dissolved
★★★★ /dɪˈzɑːlvd/
adj. 溶解的

dissolved mineral 溶解矿物质
dissolved organic matter (DOM) 溶解有机物

Groundwater usually acquires more **dissolved** minerals than surface run-off.
地下水的溶解矿物质通常比地表径流更多。

Day 188

evaporation

★★★★

/ɪˌvæpəˈreɪʃn/
n. 蒸发

evaporation capacity　　　蒸发能力
evaporation loss　　　　　蒸发损失

Evaporation is a process by which precipitation is returned to the atmosphere as vapor.
蒸发是指降水以蒸汽形式回到大气的过程。
Climatic surveys usually focus on precipitation, temperature, **evaporation**, humidity, wind, etc.
气候调查通常关注降水、温度、蒸发、湿度和风等指标。

Day 189

collect

★★★★

/kəˈlekt/
v. 收集，采集

collect data　　　　　　　收集数据
collect soil sample　　　　采集土样
collect water sample　　　采集水样

You can also **collect** water in rain barrels for use in the garden.
你还可以用雨桶接水浇花。

Day 190

suspended

/sə'spendɪd/
adj. 悬浮的

suspended dust　　悬浮尘埃

Those solids that are not dissolved in wastewater are called **suspended** solids.
那些不溶于废水的固体物质称为悬浮固体。
Long-term water storage in reservoirs reduces the amount of **suspended** sediment and bacteria.
水库中长期蓄水，可以降低悬浮泥沙和细菌的含量。

Day 191

utilize

/'juːtəlaɪz/
v. 利用，应用

We must consider how best to **utilize** the resources we have.
我们必须考虑怎样充分利用现有的资源。
Hydraulic power can be **utilized** by people in a variety of different forms.
人们可以通过很多不同形式利用水力。

Day 192

powerhouse ★★★★

/ˈpaʊrˌhaʊs/
n. 发电厂，发电站

In some dams, the **powerhouse** is constructed on one flank of the dam.
有些大坝的发电站建在大坝的一侧。
To ensure good transmission efficiency, copper cables are used in the **powerhouse**.
发电站采用铜缆以保证良好的输电效率。

Day 193

section ★★★★

/ˈsekʃən/
n. 部分，段，剖面图

Pipe **sections** can be easily joined together with a coupling sleeve and rubber-ring gasket.
使用联接套筒和胶圈可以将管段轻松地连接起来。
Here's the outside view, and here is the overflow dam in **section**.
这是外观图，这是溢流坝的剖面图。

Day 194

horizontal
★★★★

/ˌhɑːrəˈzɑːntl/
adj. 水平的，地平线的

horizontal distance　　　　　　　水平距离

The groove has a **horizontal** bottom and vertical sides.
凹槽的底部是水平的，侧面是垂直的。
You can also insert the vertical or **horizontal** pipes first and then insert the components into the pipes, one by one.
也可以先将竖管或横管插入，然后再将构件逐个插到管子上。

Day 195

bridge
★★★★

/brɪdʒ/
n. 桥

bridge site survey　　　　　　　桥渡勘测

All hydraulic structures, be it a dam, a wharf or a **bridge**, must be very durable and reliable.
所有水工建筑物，不论是大坝、码头还是桥梁，都必须坚固耐用、安全可靠。
Bridge elevation should also be measured during the operation stage.
在运营阶段也应测量桥梁高程。

Day 196

watercourse
★★★★
/ˈwɔːtəkɔːrs/
n. 水道，河道

artificial **watercourse**   人工水道
natural **watercourse**   天然水道

Surface run-off finds its way to the **watercourse** far quicker than groundwater, so there is a higher risk of floods.
地表径流进入水道的速度远比地下水快，因此洪水风险也更大。
On its way to the **watercourse**, drainage water often flows across farmland.
在注入水道的途中，排水常常流经农田。

gauge
★★★★
/ɡeɪdʒ/
n. 测量仪器

rain **gauge**   雨量计
staff **gauge**   标尺，水尺

All **gauges** at the site will be used in accordance with the procedures outlined above.
现场的所有测量仪将按照步骤进行测量。
Staff **gauges** must be plainly visible in the photographs.
图像中的水尺必须清晰可见。

outlet
★★★★

/ˈaʊtlet/
n. 出口，排水口，泄水孔

bottom **outlet** 泄水底孔
drainage **outlet** 排水出口

The grass stabilizes the soil while providing an **outlet** for drainage.
草地在稳固土壤的同时，还提供了一个排水出口。

Note
outlet（出口）与 inlet（进口）相对。

Day 199

conduit
★★★★

/ˈkɑːndʊɪt/
n. 管道，渠道

buried **conduit** 上埋式管道
intake **conduit** 进水管道

Sometimes it is convenient or economical to adopt open channels partly or wholly as main **conduits**.
有时，为了方便或经济起见，部分或全部采用明渠作为主管道。

Day 200

第三部分
三星词汇

reduce ★★★
/rɪˈduːs/
v. 减少，缩减

Day 201

Moist and frozen soils greatly **reduce** the infiltration of meltwater and thus increase spring run-off.
潮湿和冻结的土壤极大减少了融雪下渗，从而增加了春季径流。

corrosion ★★★
/kəˈroʊʒən/
n. 腐蚀，锈蚀

Day 202

corrosion resistance　　耐蚀性
galvanic **corrosion**　　电偶腐蚀

affect ★★★
/əˈfekt/
v. 影响

Day 203

Regional drought can severely **affect** the economy and even lead to starvation in some regions.
区域性干旱会严重影响经济，甚至在一些地区引发饥荒。

steel
★★★
/stiːl/
n. 钢，钢材

steel pipe — 钢管
stainless **steel** — 不锈钢

Day 204

drought
★★★
/draʊt/
n. 干旱，旱灾

Massive erosion occurs during extreme climatic situations like **drought** and tsunami.
在干旱、海啸等极端气候条件下，会发生大规模的侵蚀。

Day 205

carry
★★★
/ˈkæri/
v. 运送，携带

Rain drops **carry** tiny dust particles and other substances.
雨滴携带微小的灰尘颗粒和其他物质。

Day 206

silt ★★★
/sɪlt/
n. 泥沙，粉砂

The accumulated **silt** has raised the river bed and made it higher than the ground, so the Yellow River's lower reaches are also called "the Suspended River".
堆积的泥沙抬高了河床，使其高于地面，故黄河下游也称"悬河"。

Day 207

temperature ★★★
/ˈtemprətʃər/
n. 温度

operating **temperature**   工作温度，操作温度
soil **temperature**   土壤温度

Note
温度的单位有摄氏度（℃）和华氏度（℉），换算关系为：
华氏度 = 32 + 摄氏度×1.8

Day 208

modeling ★★★
/ˈmɑːdlɪŋ/
n. 建模

statistical **modeling**   统计建模
system **modeling**   系统建模

Day 209

characteristic
★★★
/ˌkærəktəˈrɪstɪk/
n. 特征，特性

basin **characteristic** 流域特征
physical **characteristic** 物理特征

Day 210

ground
★★★
/ɡraʊnd/
n. 地面，土地

The **ground** was frozen solid.
大地冻实了。

Day 211

pond
★★★
/pɑːnd/
n. 池塘，水池

fish **pond** 鱼塘
retention **pond** 澄清池

Day 212

integrated
★★★

/ˈɪntɪgreɪtɪd/
adj. 综合的，完整的，集成的

Water resources assessment is becoming more essential as a key component of **Integrated** Water Resources Management.
水资源评价日益成为水资源综合治理的一个重要组成部分。

Day 213

removal
★★★

/rɪˈmuːvəl/
n. 移去，清除

Land drainage is the **removal** of excess surface and subsurface water.
土地排水指去除多余的地表水和地下水。

Day 214

volume
★★★

/ˈvɑːljəm/
n. 体积，容积，量

run-off **volume**　　　　　径流量
total **volume**　　　　　　总体积，总容积

Day 215

development
/dɪˈveləpmənt/
n. 开发，发展

development zone 开发区
water resources **development** 水资源开发

Day 216

iron
/ˈaɪərn/
n. 铁，铁离子

cast **iron** 生铁，铸铁
ductile **iron** 可锻铸铁

Day 217

complete
/kəmˈpliːt/
v. 完成

In most cases, cable cranes will be totally deconstructed once the project is **completed**.
在大多数情况下，一旦项目完成，缆索起重机就会被完全拆除。

Day 218

governor
★★★

/ˈgʌvərnər/
n. 调速器，调节器

hydraulic turbine **governor** 水轮机调速器
pressure **governor** 压力调节器

Day 219

municipal
★★★

/mjʊˈnɪsəpəl/
adj. 城市的，市政的

In the past, **municipal** and industrial sewage was a major source of pollution in streams and lakes.
过去，城市和工业污水是溪流和湖泊的主要污染源之一。

Day 220

liquid
★★★

/ˈlɪkwɪd/
adj. 液态的

liquid metal 液态金属
liquid water 液态水

Day 221

machine
★★★

/məˈʃiːn/
n. 机械，机器

hydraulic **machine** 　　　　　液压机
perpetual motion **machine** 　永动机

Day 222

clay
★★★

/kleɪ/
n. 黏土

clay pipe 　陶土管
soft **clay** 　软黏土

Day 223

safe
★★★

/seɪf/
adj. 安全的

Public drinking water systems provide **safe** drinking water for communities by using various methods of water treatment.
公共饮用水系统使用各种水处理方法，为社区提供安全饮用水。

Day 224

motor
★★★
/ˈmoʊtər/
n. 马达，发动机，电动机

electric **motor** 电动机
hydraulic **motor** 液压马达

Day 225

sensor
★★★
/ˈsensər/
n. 传感器

pressure **sensor** 压力传感器
soil moisture **sensor** 土壤湿度传感器

Day 226

base
★★★
/beɪs/
n. 基底，基础

Base flow is water that flows into the stream from natural storage.
基流是从自然储水流入河流中的水。

Day 227

hardness
★★★

/ˈhɑːrdnɪs/
n. 硬度

hardness test 硬度试验
carbonate **hardness** 碳酸盐硬度
water **hardness** 水硬度

Day 228

underground
★★★

/ˌʌndərˈɡraʊnd/
adj. 位于地下的

underground reservoir 地下水库
underground water 地下水

Day 229

tillage
★★★

/ˈtɪlɪdʒ/
n. 耕作，耕种

conventional **tillage** 传统耕作
strip **tillage** 带状耕作

Day 230

speed
★★★
/spiːd/
n. 速度，速率

rated **speed** 额定速度
rotational **speed** 转速

Day 231

current
★★★
/ˈkɜːrənt/
n. （水，气，电）流

alternating **current** 交流电
direct **current** 直流电

Day 232

circuit
★★★
/ˈsɜːrkɪt/
n. 电路，回路

The protection device shall be sized appropriately to protect against overload and short **circuits**.
保护装置应该大小合适，以防止过载和短路。

Day 233

output ★★★
/ˈaʊtpʊt/
n. 产量，输出功率

agricultural **output** — 农业产量
net **output** — 净输出，净产量

Day 234

implementation ★★★
/ˌɪmpləmenˈteɪʃn/
n. 实施，执行，实现

Project costs and time requirements may be updated during project **implementation**.
项目成本和时间要求可能会在项目实施期间更新。

Day 235

diversion ★★★
/daɪˈvɜːrʒən/
n. 分水，引流

diversion dam — 分水坝
diversion structure — 导流建筑物

Day 236

navigation ★★★
/ˌnævəˈgeɪʃən/
n. 航运，航行，航海

The advantage of an open channel is that it could be used for irrigation or **navigation** purposes.
明渠的优点是其可以用于灌溉或航运。

Day 237

seal ★★★
/siːl/
v. 把……封住，封闭

The casings of the well should be **sealed** with grout or other cement material.
井的套管应该用水泥薄浆或其他胶泥材料密封。

Day 238

bucket ★★★
/ˈbʌkɪt/
n. 桶，铲斗

bucket elevator dredger   链斗式挖泥船

Note
链斗式挖泥船指用装在斗桥上的系列链斗，连续循环运转而进行疏浚的挖泥船。

Day 239

pollutant
★★★

/pəˈluːtənt/
n. 污染物

All kinds of harmful substances, pesticides, and other **pollutants** can cause pollution to drinking water sources.
各种有害物质、农药及其他污染物都可能污染饮用水水源。

Day 240

port
★★★

/pɔːrt/
n. 港口，口岸

According to the location, the **ports** can be classified into lake **ports**, river **ports** and sea **ports**, etc.
根据地理位置，港口可分为湖港、河港和海港等。

Day 241

civil
★★★

/ˈsɪvəl/
adj. 公民的，民用的，民事的

civil architecture	民用建筑
civil engineering	土木工程

Day 242

meter
★★★

/ˈmiːtər/
n. 计，表，米（单位）

cubic **meter** 立方米
current **meter** 流速仪

Day 243

evaluate
★★★

/ɪˈvæljueɪt/
v. 评价，估价

It can be difficult to **evaluate** the effectiveness of different methods.
很难对各种方法的效果作出评估。

Day 244

cycle
★★★

/ˈsaɪkəl/
n. 循环，周期

life **cycle** 生命周期，使用周期
water **cycle** 水循环

Day 245

ditch
★★★

/dɪtʃ/
n. 沟渠

Ditches include sewage pipes and rainwater sewers that can be seen everywhere on the roadside.
沟渠包含污水管道及路边随处可见的雨水下水道。

Day 246

culvert
★★★

/ˈkʌlvərt/
n. 涵洞，排水道

arch **culvert**   拱形涵洞
circular **culvert**   圆形涵洞

Day 247

sludge
★★★

/slʌdʒ/
n. 污泥，淤泥

The secondary clarifier removes the activated **sludge** from water.
二次澄清池去除水中的活性污泥。

Day 248

wetland
★★★

/ˈwetlənd/
n. 湿地，沼泽地

wetland run-off 湿地径流
artificial wetland 人工湿地

Day 249

vary
★★★

/ˈveri/
v. 变化

The flow that does not **vary** with time is called the steady flow.
不随时间变化的水流称为恒流。

Day 250

infrastructure
★★★

/ˈɪnfrəˌstrʌktʃər/
n. 基础设施

Infrastructures like community hospitals are intended for reducing some waterborne diseases.
社区医院等基础设施旨在减少某些水传疾病。

Day 251

vegetation
★★★
/ˌvedʒɪˈteɪʃən/
n. 植被，植物

Soil erosion by wind is due to lack of **vegetation** cover.
土壤被风侵蚀是植被覆盖不足造成的。

Day 252

preparation
★★★
/ˌprepəˈreɪʃən/
n. 预备，准备

Preparations for the hydropower station are nearing completion.
建设水电站的前期准备正接近尾声。

Day 253

gravity
★★★
/ˈɡrævəti/
n. 重力，地心引力

gravity dam　　　　　　　重力坝
gravity irrigation　　　　　自留灌溉

Day 254

surge
★★★

/sɜːrdʒ/
n. (电流)浪涌

surge voltage   浪涌电压
electrical **surge**   电涌

Day 255

excess
★★★

/ˈekses/
adj. 额外的，过量的

There should be a guard against **excess** power supply in any electric wiring system.
在任何电力布线系统中，都应该有一个防止过度供电的装置。

Day 256

breakwater
★★★

/ˈbreɪkˌwɔːtər/
n. 防波堤

floating **breakwater**   浮式防波堤
submerged **breakwater**   水下防波堤

Day 257

rainwater
★★★

/ˈreɪnwɔːtər/
n. 雨水

The professional drainage engineer is responsible for directing **rainwater** run-off from highways and roads.
这位专业排水工程师负责主路和小路上的雨水引流。

Day 258

impact
★★★

/ɪmˈpækt/
v. 影响

If wastewater is not properly treated, the environment and human health can be negatively **impacted**.
污水如果处理不当，会对环境和人类健康产生负面影响。

Day 259

vessel
★★★

/ˈvesəl/
n. 船，舰，容器

container **vessel**	集装箱船
pressure **vessel**	压力容器

Day 260

effluent
★★★

/ˈefluənt/
n. 污水

In some cases, the **effluent** resulting from secondary treatment is not clean enough for discharge.
有时候，二级处理后产生的污水不够干净，达不到排放标准。

Day 261

guideline
★★★

/ˈgaɪdlaɪn/
n. 指导方针，参考

safety **guideline**   安全指南

The WHO has issued **guidelines** for drinking water quality.
世界卫生组织发布了饮用水质量指南。

Day 262

auxiliary
★★★

/ɔːgˈzɪljəri/
adj. 辅助的，备用的

auxiliary equipment   辅助设备
auxiliary system   辅助系统

Day 263

renewable
★★★

/rɪˈnuːəbəl/
adj. 可再生的，可更新的

renewable energy 可再生能源
renewable resource 可再生资源

Day 264

seepage
★★★

/ˈsiːpɪdʒ/
n. 渗漏

seepage control 防渗
seepage discharge 渗流量
seepage loss 渗漏损失

Day 265

bottom
★★★

/ˈbɑːtəm/
n. 底部

The layer of concentrated solids that is collected at the **bottom** of the tank is called sludge.
在沉淀池底部收集的浓缩固体层被称为淤泥。

Day 266

personnel
★★★

/ˌpɜːrsəˈnel/
n. 全体人员，员工

maintenance **personnel** — 维修人员
military **personnel** — 军事人员
technical **personnel** — 技术人员

Day 267

contractor
★★★

/ˈkɑːntræktər/
n. 承包人，承包商

Costs for equipment can vary widely among **contractors**.
不同承包商的设备成本差别较大。

Day 268

artificial
★★★

/ˌɑːrtɪˈfɪʃəl/
adj. 人造的，人工的

artificial fertilizer — 人造肥料
artificial intelligence — 人工智能

Day 269

layer
★★★
/ˈleɪər/
n. 层,表层

It's necessary to compact the soil **layers** under the pipe so as to provide proper support.
管道下面的土层必须压实,以提供适当的支撑。

Day 270

physical
★★★
/ˈfɪzɪkəl/
adj. 物理的

Sewage treatment includes **physical** treatment, chemical treatment and biological treatment.
污水处理包括物理法、化学法和生物法。

Day 271

hazard
★★★
/ˈhæzərd/
n. 危险,危害

natural **hazard**　　　　　　　自然灾害

Long-term exposure to stone powder will be a **hazard** to the human body.
长期接触滑石粉会对人体造成危害。

Day 272

procedure
★★★

/prəˈsiːdʒər/
n. 程序，步骤

operating **procedure** 操作程序
testing **procedure** 测试程序

Day 273

microorganism
★★★

/ˌmaɪkroʊˈɔːrɡənɪzəm/
n. 微生物

Water can support the growth of many types of **microorganisms**.
水可以供养多种微生物。

Day 274

calculate
★★★

/ˈkælkjəleɪt/
v. 计算

From this you can **calculate** the total mass of the gravity dam.
你可以据此计算出重力坝的总重量。

Day 275

check ★★★
/tʃek/
n.&v. 检查，控制

- check out —— 查看，检查
- check valve —— 止回阀
- security check —— 安全检查

Day 276

damage ★★★
/ˈdæmɪdʒ/
n. 损害，破坏

- cavitation damage —— 空蚀
- environmental damage —— 环境破坏

Day 277

sample ★★★
/ˈsæmpəl/
n. 样品，样本

Hydrologists may collect **samples** of soil, rock and water for laboratory analysis.
水文工作者可能会收集土、岩石和水的样本进行实验室分析。

Day 278

cylinder
★★★
/ˈsɪləndər/
n. 汽缸

double-acting **cylinder** 双作用缸
hydraulic **cylinder** 液压缸

Day 279

demand
★★★
/dɪˈmænd/
n. 需求，需求量

electricity **demand** 电力需求
supply and **demand** 供需

Day 280

algae
★★★
/ˈælgiː/
n. 藻类，海藻

The overgrowth of **algae** can alter the pH of water.
藻类的过度生长能改变水的 pH 值。

Day 281

duration
★★★

/dʊˈreɪʃən/
n. 历时，持续时间

Total run-off for a storm is related to the rainfall **duration** and intensity.
暴雨的总径流量与降雨历时及强度有关。

Day 282

reclamation
★★★

/ˌrekləˈmeɪʃən/
n. 开垦，回收

land **reclamation**   垦荒
wastewater **reclamation**   废水回收

Day 283

manual
★★★

/ˈmænjuəl/
adj. 手动的
n. 使用手册

manual control   手动控制
operation **manual**   操作手册
reference **manual**   参考手册

Day 284

urban
★★★
/ˈɜːrbən/
adj. 城市的，市区的

urban area　　　　　　　　城市地区
urban drainage　　　　　　城市排水

Day 285

continuous
★★★
/kənˈtɪnjuəs/
adj. 持续的，连续的

Direct and **continuous** monitoring of the flow cannot be carried out except for very specific cases.
除非极特殊情况，否则不能对流量进行直接、连续的监测。

Day 286

remote
★★★
/rɪˈmoʊt/
adj. 远程的

Water resources assessment uses **remote** sensing for analysis.
水资源评估运用遥感技术进行分析。

Day 287

automation
★★★

/ˌɔːtəˈmeɪʃən/
n. 自动化

The control and **automation** systems are critical to high efficiency in operation.
控制和自动化系统是高效运行的关键。

Day 288

potential
★★★

/pəˈtenʃəl/
n. 势能

The greater the height of water, the more **potential** energy it contains.
水的高度越高，势能就越大。

Day 289

connect
★★★

/kəˈnekt/
v. 连接

Cables **connected** to various pieces of equipment should be double insulated.
连接各种设备的电缆必须双重绝缘。

Day 290

frequency
★★★

/ˈfriːkwənsi/
n. 频率

Day 291

Climate change will worsen the situation by increasing the **frequency** and intensity of floods and droughts.
气候变化将导致旱涝灾害的频率及强度增加，使情况恶化。

barrage
★★★

/ˈbɑːrɪdʒ/
n. 拦河坝

Day 292

Barrages are typical water retaining structures.
拦河坝是典型的挡水建筑物。

odor
★★★

/ˈoʊdər/
n. 气味

Day 293

Odor and taste are associated with the presence of living microorganisms.
气味和味道与活的微生物有关。

adequate
★★★

/ˈædɪkwət/
adj. 充足的，足够的

Water distribution systems are built to provide **adequate** water pressure and flow.
建造配水系统是为了提供足够的水压和水流。

Day 294

recharge
★★★

/ˌriːˈtʃɑːrdʒ/
n. 回灌

If the final effluent is sufficiently clean, it can also be used for groundwater **recharge**.
如果最后排出的污水足够干净，也可以用于地下水回灌。

Day 295

rotor
★★★

/ˈroʊtər/
n. 转子，旋转体

rotor winding　　　　　　　　转子绕组
generator **rotor**　　　　　　　发电机转子

Day 296

bid
★★★

/bɪd/
n.&v. 出价，投标

bid bond　　投标保证金

Three firms **bid** for the contract on the new dams.
有三家公司竞标承包新大坝工程。

Day 297

compliance
★★★

/kəmˈplaɪəns/
n. 服从，遵守

Compliance with terms in the contract is essential in the development of the project.
遵守合同条款对该项目的发展至关重要。

Day 298

disposal
★★★

/dɪˈspoʊzəl/
n. 处理

Hydrologists provide guidance on the location of monitoring wells around waste **disposal** sites.
水文学者为废物处理场周围监测井的选址提供指导。

Day 299

stage ★★★
/steɪdʒ/
n. 水位，阶段

A flood warning will be issued when river or lake levels exceed (or are expected to exceed) the flood **stage** within the next 24 hours.
当河流或湖泊水位在未来 24 小时内超过或预计将超过洪水位时，会发出洪水警告。

Day 300

sprinkler ★★★
/ˈsprɪŋklər/
n. 喷灌机，洒水器

Alarms went off and the **sprinkler** system turned on.
警报声响，喷淋系统启动。

Day 301

rural ★★★
/ˈrʊrəl/
adj. 农村的，乡村的

rural area　　　　　　　农村地区
rural water supply　　　农村水供应

Day 302

embankment
★★★

/ɪmˈbæŋkmənt/
n. 堤岸，堤防

earth **embankment** 土堤
lake **embankment** 湖堤

Day 303

coastal
★★★

/ˈkoʊstl/
adj. 沿海的，海岸的

Coastal areas are sometimes flooded by unusually high tides.
沿海地区有时会被异常高的潮水淹没。

Day 304

essential
★★★

/ɪˈsenʃəl/
adj. 基本的，必要的

Water is **essential** for food production.
对粮食生产而言，水是必不可少的。

Day 305

efficiency
★★★ /ɪˈfɪʃənsi/
n. 效率

maximum **efficiency** — 最大效率
water use **efficiency** — 用水效率

Day 306

recreational
★★★ /ˌrekriˈeɪʃənl/
adj. 娱乐的，消遣的

recreational activity — 娱乐活动
recreational facilities — 娱乐设施

Day 307

roadway
★★★ /ˈroʊdweɪ/
n. 车行道

Some prolonged high floods can delay traffic in areas which lack elevated **roadways**.
在缺乏高架路的地区，持续较久的高洪会导致交通阻延。

Day 308

aquatic
★★★

/əˈkwætɪk/
adj. 水生的，水栖的

aquatic ecosystem　　　水生生态系统
aquatic life　　　水生生物

Day 309

clearance
★★★

/ˈklɪrəns/
n. 间隙，清除

There has to be **clearance** between the turbine and the generator.
水轮机和发电机之间必须有间隙。

Day 310

manufacturer
★★★

/ˌmænjəˈfæktʃərər/
n. 生产商，制造商

All these instruments are available from any of the major water monitoring instrument **manufacturers**.
所有这些仪器都可以从任意一家水监测仪器大制造商那里买到。

Day 311

buoy
★★★

/ˈbuːi/
n. 浮标，航标

Buoys are of different designs and types.
浮标有不同的设计和种类。

Day 312

geological
★★★

/ˌdʒiːəˈlɑːdʒɪkəl/
adj. 地质的，地质学的

Historical and real-time data are available from the China **Geological** Survey.
历史数据和实时数据可从中国地质调查局获得。

Day 313

rotate
★★★

/ˈroʊteɪt/
v. 旋转，循环

The rotor **rotates** at fixed speed.
转子以固定的速度旋转。

Day 314

evaluation
★★★

/ɪˌvæljuˈeɪʃən/
n. 评价，评估

Monitoring and **evaluation** are essential for ensuring that the current management of water resources is properly implemented.
为了确保当前水资源管理的妥善实施，监测与评估必不可少。

Day 315

wire
★★★

/waɪr/
n. 电线，金属丝

copper **wire**   铜丝
earth **wire**   接地线

Day 316

objective
★★★

/əbˈdʒektɪv/
n. 目的，目标

The primary **objective** of water treatment is to protect the health of the community.
水处理的主要目标是保护社区健康。

Day 317

specific
★★★

/spəˈsɪfɪk/
adj. 特定的

Those water conservancy projects combine with the **specific** local landform and are very impressive.
那些水利工程与当地特定的地形地貌结合在一起，令人印象深刻。

Day 318

sufficient
★★★

/səˈfɪʃənt/
adj. 足够的，充分的

The dam is constructed on a large river in hilly areas to ensure **sufficient** water storage at height.
大坝建在位于丘陵地区的一条大河上，以确保在高处有足够的蓄水量。

Day 319

mitigation
★★★

/ˌmɪtɪˈɡeɪʃən/
n. 减轻，缓和

mitigation measure　　　　　　缓解措施
disaster **mitigation**　　　　　　减灾

Day 320

residual
★★★

/rɪˈzɪdʒuəl/
adj. 剩余的，残留的
n. 剩余物

residual value　　残值

It leaves no **residual**, and it does not cause taste or odor problems.
它不会有任何残留，也不会产生味道或气味问题。

Day 321

invert
★★★

/ɪnˈvɜːrt/
n. 倒拱，管道内底
v. 使倒置

invert elevation　　管内底标高
inverted siphon　　倒虹吸管

Day 322

line
★★★

/laɪn/
n. 线，线路

The transmission **lines** can be extended to the local village to supply domestic power.
输电线路可以延伸至当地村庄，提供家用电。

Day 323

real-time
★★★
/ˈrɪəl taɪm/
adj. 实时的

real-**time** data — 实时数据
real-**time** monitoring — 实时监测

Day 324

zone
★★★
/zoʊn/
n. 地区，地带

Frequent and light irrigation helps keep water and mobile nutrients in the root **zone**.
频繁的浅灌有助于将水分和流动的养分留在根区。

Day 325

compact
★★★
/kəmˈpækt/
v. 压实，夯实

Rollers are driven over the concrete to **compact** it down.
压路机碾过混凝土，将其压实。

Day 326

bearing
★★★

/ˈberɪŋ/
n. 轴承，方位

bearing temperature 轴承温度
ball **bearing** 滚珠轴承

Day 327

economical
★★★

/ˌekəˈnɑːmɪkəl/
adj. 经济实惠的，节约的

It should be **economical** to load the vessels when the water level is low.
当水位较低时装载船只，应该是经济实惠的。

Note
注意：economic 的意思为"经济的，经济学的"。

Day 328

cement
★★★

/sɪˈment/
n. 水泥

cement content 水泥含量
cement grouting 水泥灌浆

Day 329

generation
★★★

/ˌdʒenəˈreɪʃən/
n. (电、热能的)产生

Day 330

hydropower **generation**   水力发电

From the viewpoint of power **generation**, water is obviously a resource.
从发电的角度来看，水显然是一种资源。

criterion
★★★

/kraɪˈtɪriən/
n. 标准，准则

Day 331

design **criterion**   设计准则
drainage **criterion**   排水标准

Note
该词的复数形式为 criteria。

shutdown
★★★

/ˈʃʌtdaʊn/
n. 关机，停工

Day 332

The abnormal **shutdown** prevents the device from being managed by another operator.
异常关闭使另一位管理员无法对设备进行管理。

impurity
★★★

/ɪmˈpjʊrəti/
n. 杂质

Day 333

A water filter is a device which removes **impurities** from water.
滤水器是一种去除水中杂质的装置。

pipeline
★★★

/ˈpaɪplaɪn/
n. 管道

Day 334

A water distribution **pipeline** must be able to resist internal and external forces.
配水管道必须能够承受内力和外力。

service
★★★

/ˈsɜːrvɪs/
n. 服务，检修

Day 335

service life 使用寿命，使用年限
water **service** 供水

That machine is in for a **service**.
那台机器送去检修了。

instrument
★★★

/ˈɪnstrəmənt/
n. 仪器，仪表

measuring **instrument** 计量仪器
monitoring **instrument** 监测仪器

Day 336

float
★★★

/fləʊt/
v. 漂浮，浮动
n. 浮标，浮子

floating debris 漂浮物
surface **float** 水面浮标

The **float** should not touch the ground.
浮子不应接触地面。

Day 337

yield
★★★

/jiːld/
n. 产量

sediment **yield** 产沙量

Increased crop **yields** can be attained within the first year.
第一年就可以实现作物增产。

Day 338

outflow ★★★
/ˈaʊtfloʊ/
n. 流出，流量

outflow channel 放水渠道
outflow rate 流出量

Day 339

excavator ★★★
/ˈekskəveɪtər/
n. 挖掘机

dragline excavator 拉铲挖掘机
multi-bucket excavator 多斗挖掘机
single-bucket excavator 单斗挖掘机

Day 340

tube ★★★
/tuːb/
n. 管

tube well 管井
draft tube 尾水管

Day 341

hazardous
★★★

/ˈhæzərdəs/
adj. 有害的，危险的

hazardous to health 对健康有害

The disposal of **hazardous** waste is of vital importance.
有害废弃物的处理至关重要。

Day 342

drawing
★★★

/ˈdrɔːɪŋ/
n. 图纸，绘图

drawing board 制图板
drawing template 绘图模板
construction **drawing** 施工图，结构图
engineering **drawing** 工程制图

Day 343

hose
★★★

/hoʊz/
n. 软管

hose reel 软管卷
rubber **hose** 橡胶软管

Day 344

transmission
★★★

/trænsˈmɪʃən/
n. 输电，输送

A **transmission** line is used for the **transmission** of electric power from power stations and substations to the various distribution units.
输电线路用来将电力从发电站和变电站输送到各配电单元。

Day 345

sustainable
★★★

/səˈsteɪnəbəl/
adj. 可持续的

sustainable development　　可持续发展

Soil moisture content is a basic parameter for studying **sustainable** management of water.
土壤含水量是研究水资源可持续管理的一项基本参数。

Day 346

cable
★★★

/ˈkeɪbəl/
n. 缆索，电缆

Cable crane systems have a high level of safety which can avoid common risks.
缆索起重机系统的安全等级较高，可以避免常见的风险。

Day 347

diagram
★★★

/ˈdaɪəɡræm/
n. 图解，示意图

block **diagram**	框图，块状图
circuit **diagram**	电路图
wiring **diagram**	布线图

Day 348

substance
★★★

/ˈsʌbstəns/
n. 物质

harmful **substance**	有害物质
organic **substance**	有机物质
toxic **substance**	有毒物质

Day 349

climatic
★★★

/klaɪˈmætɪk/
adj. 气候的

The principal **climatic** factors that affect run-off include precipitation and evaporation.
影响径流的主要气候因素包括降水和蒸发。

Day 350

curve
★★★

/kɜːrv/
n. 曲线

demand **curve** 需求曲线
frequency **curve** 频率曲线
gravity **curve** 抛物线

Day 351

earth
★★★

/ɜːrθ/
n. 地球，陆地，土壤

Water covers more than two-thirds of the **Earth**'s surface.
水覆盖了地球表面的三分之二以上。

Day 352

loss
★★★

/lɒːs/
n. 损失，损耗，亏损

In reservoirs, seepage is the main issue behind water **losses**.
在水库中，渗流是造成水流失的关键。

Day 353

rotation
★★★

/roʊˈteɪʃən/
n. 轮作，旋转

Crop **rotation** is the successive cultivation of different crops in a specified order on the same field.
农作物轮作指在同一块土地上按一定的顺序接连种植不同的农作物。

Day 354

fault
★★★

/fɒːlt/
n. 故障，错误

The **fault** in electrical equipment will cause damage and disturb the normal flow of the electric current.
电气设备的故障会造成损坏，干扰电流的正常流动。

Day 355

grid
★★★

/grɪd/
n. 网格，电力网

grid connection　　　　　　　栅极接线
intelligent power **grid**　　　　智能电网
State **Grid** Corporation of China　国家电网

Day 356

rain
★★★

/reɪn/
n. 雨，降雨

Precipitation is the water falling from the cloud in the form of snow, **rain**, hail, and sleet.
降水是指云中的水以雪、雨、冰雹和雨夹雪的形式降落到地面。

Day 357

microbiology
★★★

/ˌmaɪkroʊbaɪˈɑːlədʒi/
n. 微生物学

Microbiology is the study of microscopic organisms, such as bacteria, fungi, and protozoa.
微生物学是研究微生物，如细菌、真菌和原生动物的学科。

Day 358

wet
★★★

/wet/
adj. 湿的，潮湿的

Dry areas are more prone to erosion than **wet** areas.
干燥地区比潮湿地区更容易受到侵蚀。

Day 359

masonry
★★★

/ˈmeɪsənri/
n. 砖石，砌石技艺

brick **masonry** structure 砖砌结构
concrete **masonry** 混凝土砌体

Day 360

weld
★★★

/weld/
v. 焊接

Sections of a steel pipe are joined by **welding** or mechanical coupling devices.
钢管管段之间可以通过焊接或机械耦合装置连接在一起。

Day 361

suction
★★★

/ˈsʌkʃən/
n. 吸，抽吸

The **suction** filter is a self-sealing filter like a valve which can operate automatically when the canister is full.
吸滤器是一种像阀门一样的自密封过滤器，在滤罐变满后自动开始工作。

Day 362

inflow
★★★
/ˈɪnfloʊ/
n. 流入

inflow rate — 入流量，进水率
surface inflow — 地表入流

Day 363

hammer
★★★
/ˈhæmər/
n. 锤子

hammer ram — 夯锤
water hammer — 水锤，水击

Day 364

intensity
★★★
/ɪnˈtensəti/
n. 强度，密度

run-off intensity — 径流强度

Day 365

附录一　音标表

本书读音以国际音标（International Phonetic Alphabet，IPA）标示。

Consonants 辅音

Symbol 符号	keyword 示例
p	pen
b	back
t	ten
d	day
k	key
g	get
f	fat
v	view
θ	thing
ð	then
s	soon
z	zero
ʃ	ship
ʒ	pleasure
h	hot
x	loch
tʃ	cheer
dʒ	jump
m	sum
n	sun
ŋ	sung
w	wet
l	let
r	red
j	yet

Vowels 元音

	Symbol 符号	Keyword 示例
short 短元音	ɪ	bit
	e	bed
	æ	cat
	ɒ	dog (*BrE*【英】)
	ʌ	cut
	ʊ	put
	ə	about
	i	happy
	u	actuality
long 长元音	iː	sheep
	ɑː	father
	ɒː	dog (*AmE*【美】)
	ɔː	four
	uː	boot
	ɜː	bird
diphthongs 双元音	eɪ	make
	aɪ	lie
	ɔɪ	boy
	əʊ	note (*BrE*【英】)
	oʊ	note (*AmE*【美】)
	aʊ	now
	ɪə	real
	eə	hair (*BrE*【英】)
	ʊə	tour (*BrE*【英】)

Special signs 特殊符号

/ˈ/	表示主重音
/ˌ/	表示次重音
/ə/	表示 /ə/ 可发音可不发音

附录二　总词表

A

adequate	191
affect	160
agricultural	131
algae	186
analysis	137
application	132
aquatic	196
aquifer	138
area	33
artificial	182
assessment	123
automation	189
auxiliary	180
available	111

B

bacteria	118
barrage	190
base	168
basin	63
bearing	202
bid	192
bottom	181
breaker	142
breakwater	178
bridge	156
bucket	172
buoy	197

C

cable	208
calculate	184
canal	30
capacity	85
carry	161
catchment	106
cause	133
cement	202
channel	81
characteristic	163
check	185
chemical	67
circuit	170
civil	173
clay	167
clearance	196
climatic	209
coastal	194
collect	153
compact	201
complete	165

compliance	192		demand	186
component	56		depend	134
concrete	39		depth	68
condition	74		design	11
conduit	158		development	165
connect	189		device	147
conservation	36		diagram	209
construction	15		discharge	31
contaminant	129		disinfection	130
contamination	122		disposal	192
continuous	188		dissolved	152
contour	139		distribution	90
contractor	182		district	151
control	24		ditch	175
corrosion	160		diversion	171
cost	40		downstream	94
crane	135		drain	86
criterion	203		drainage	12
crop	69		drawing	207
cross-section	144		drip	143
culvert	175		drought	161
current	170		duration	187
curve	210		**E**	
cycle	174		earth	210
cylinder	186		economical	202
D			efficiency	195
dam	8		effluent	180
damage	185		electrical	99
data	34		electricity	101

elevation	60
embankment	194
energy	124
engineering	25
ensure	120
environmental	57
equipment	9
erosion	43
essential	194
estimate	150
evaluate	174
evaluation	198
evaporation	153
excavator	206
excess	178

F

facility	120
factor	95
fault	211
field	73
filter	66
filtration	109
float	205
flood	20
flow	5
fluid	59
forecast	130
frequency	190

G

gate	133
gauge	157
generation	203
generator	22
geological	197
governor	166
gravity	177
grid	211
ground	163
groundwater	32
guideline	180

H

hammer	214
harbor	121
hardness	169
hazard	183
hazardous	207
horizontal	156
hose	207
hydraulic	3
hydro-	41
hydroelectric	53
hydrograph	96
hydrologic	72
hydrology	44
hydropower	19

I

impact	179

implementation	171
impurity	204
include	112
industrial	127
infiltration	91
inflow	214
information	127
infrastructure	176
inlet	128
inspection	126
install	79
installation	113
instrument	205
intake	152
integrated	164
intensity	214
invert	200
iron	165
irrigation	16

L

lake	110
land	80
layer	183
level	71
line	200
liquid	166
load	65
location	87
loss	210

M

machine	167
machinery	145
maintain	151
maintenance	46
management	50
manual	187
manufacturer	196
map	150
masonry	213
material	54
maximum	146
measure	98
measurement	117
mechanical	105
meter	174
method	45
microbiology	212
microorganism	184
minimum	117
mitigation	199
modeling	162
moisture	100
monitoring	35
motor	168
municipal	166

N

natural	93
navigation	172

necessary	118	power	37
normal	116	powerhouse	155

O

objective	198	precipitation	51
obtain	138	preparation	177
odor	190	pressure	62
operate	115	prevent	149
operation	78	procedure	184
organic	126	process	115
outflow	206	profile	141
outlet	158	project	21
output	171	protection	108
oxygen	142	provide	82
		pump	17

P

Q

particle	148	quality	26
peak	114	quantity	116
penstock	97		

R

perform	145	rain	212
personnel	182	rainfall	64
physical	183	rainwater	179
pipe	23	rate	122
pipeline	204	real-time	201
plan	143	recharge	191
plant	13	reclamation	187
pollutant	173	recreational	195
pollution	104	reduce	160
pond	163	relay	128
port	173	remote	188
potential	189	removal	164

remove	131	silt	162
renewable	181	siphon	146
repair	124	site	77
requirement	107	size	148
reservoir	27	slope	61
residual	200	sludge	175
resource	29	soil	4
riparian	136	solid	125
river	10	solution	140
roadway	195	source	70
rotate	197	specific	199
rotation	211	speed	170
rotor	191	spillway	129
run-off	14	sprinkler	193
rural	193	stage	193
		standard	141

S

safe	167	station	76
sample	185	steel	161
sand	136	storage	49
seal	172	storm	137
section	155	stream	58
sediment	89	structure	7
seepage	181	substance	209
sensor	168	subsurface	147
service	204	suction	213
sewage	83	sufficient	199
sewer	134	suitable	144
shaft	139	supply	18
shutdown	203	surface	28

surge	178	vary	176
survey	92	vegetation	177
suspended	154	velocity	84
sustainable	208	vertical	123

T

tank	75	vessel	179
technical	140	voltage	119
temperature	162	volume	164

W

test	55	waste	149
tillage	169	wastewater	48
transformer	135	water	2
transmission	208	watercourse	157
transport	125	watershed	38
treatment	42	waterway	88
tube	206	weir	114
turbidity	132	weld	213
turbine	6	well	119

U

		wet	212
underground	169	wetland	176
unit	47	wire	198

Y

upstream	121		
urban	188	yield	205
utilize	154		

Z

		zone	201

V

valve 52